VEHICULAR SOCIAL NETWORKS

VEHICULAR SOCIAL NETWORKS

Edited by
Anna Maria Vegni
Valeria Loscrí
Athanasios V. Vasilakos

CRC Press
Taylor & Francis Group
Boca Raton London New York

CRC Press is an imprint of the
Taylor & Francis Group, an **informa** business

CRC Press
Taylor & Francis Group
6000 Broken Sound Parkway NW, Suite 300
Boca Raton, FL 33487-2742

First issued in paperback 2020

© 2017 by Taylor & Francis Group, LLC
CRC Press is an imprint of Taylor & Francis Group, an Informa business

No claim to original U.S. Government works

ISBN 13: 978-0-367-57380-5 (pbk)
ISBN 13: 978-1-4987-4919-0 (hbk)

Library of Congress Cataloging-in-Publication Data

Names: Vegni, Anna Maria. | Loscrí, Valeria. | Vasilakos, Athanasios.
Title: Vehicular social networks / edited by Anna Maria Vegni, Valeria Loscrí and Athanasios V. Vasilakos.
Description: Boca Raton, FL : CRC Press, Taylor & Francis Group, [2017] | Includes bibliographical references and index.
Identifiers: LCCN 2016028987 | ISBN 9781498749190 (hardback : acid-free paper) | ISBN 9781498749206 (ebook)
Subjects: LCSH: Vehicular ad hoc networks (Computer networks) | Online social networks.
Classification: LCC TE228.37 .V49 2017 | DDC 388.3/12–dc23
LC record available at https://lccn.loc.gov/2016028987

Visit the Taylor & Francis Web site at
http://www.taylorandfrancis.com

and the CRC Press Web site at
http://www.crcpress.com

To my daughter, Giulia

-Anna Maria Vegni

To my children, Gabriele and Elena

-Valeria Loscrí

To my son, Vasileios

-Athanasios V. Vasilakos

Contents

Preface

Overview and Goals

Cars, and vehicles in general, have changed significantly over the years and will continue to do so in the future. In particular, the integration of more and more sensors, such as cameras and radar, and communication technologies opens up a whole new design space for in-vehicle applications. It is expected that a different kind of automotive experience will emerge, where city streets will teem with small, driverless cars whose wireless capabilities will direct traffic flow smoothly, rendering traffic lights unnecessary. Furthermore, the use of cloud computing technology will enable passengers to work or play games during their commutes, while listening to their favorite music, as chosen by the car based on user profile.

Vehicular communications can be considered the "first social network for automobiles," since each driver can share data with other neighbors. For instance, heavy traffic is a common occurrence in some areas on the roads (e.g., at intersections and taxi loading and unloading areas); as a consequence, roads are becoming a popular social place for vehicles to connect to each other. Social characteristics and human behavior largely impact vehicular ad hoc networks leading to the rise of vehicular social networks (VSNs), which are formed when vehicles (individuals) "socialize" and share common interests.

The goal of this book is to provide the main features of VSNs, from novel emerging technologies to social aspects used for mobile applications, as well as the main issues and challenges. VSNs are described as decentralized opportunistic communication networks formed among vehicles. They exploit mobility aspects and the basics of traditional social networks in order to create novel approaches of message exchange through the detection of dynamic social structures. Particular attention is given to social aspects that are exploited in vehicular communications for both safety and entertainment applications.

Selected topics that are covered in this book include social networking techniques, social-based routing techniques applied to vehicular networks, data dissemination in VSNs, architectures for VSNs, and novel trends and challenges in VSNs. The book aims to provide significant technical and practical insights into different aspects, starting from a basic background on social networking and going on to cover the interrelated technologies and applications of vehicular ad hoc networks, the technical challenges, implementation, and future trends.

This book comes as a joint effort from very high-profile researchers, expert in the vehicular social domain, and each of them in specific specialties of this large research field. We tried to contribute to ongoing and emerging topics in the context of VSNs by following a pedagogical approach in order to realize a final book targeting both students and professional researchers. In particular, the different contributions, coming from different authors, have been carefully homogenized and synchronized in order to allow for easy reading for all potential readers.

Features of the Book

Some highlights of the book are as follows:

- VSNs are an emergent topic that is attracting a lot of interest in the research community. This book strives to be the first collection of different contributions and attempts to maintain continuity of character.

- The investigation of social aspects as an answer to the existing challenges is important and deserves thorough study; thus, it occupies an important place in the book.

- This book offers a study of the main features and fundamentals of VSNs, starting with an introduction of the architectures for VSNs, progressing to the logistics of effectively disseminating data in VSNs, and finally, investigating the applications in the context of VSNs by taking into account the security aspects.

Organization of Chapters

The book is organized into nine chapters, with each written by one or more topical area experts. The chapters are grouped into three parts.

Part I, "Introduction to Vehicular Social Networks," is devoted to a brief introduction of VSNs and their main features. Specifically, Chapter 1 introduces a novel paradigm, vehicular social sensor networks, to explore social sensing mechanisms in VSNs and investigate their social sensing capabilities in VSNs. Then, Chapter 2 deals with the design of architectures for VSN to cope with several challenges such as mobility, multihoming, and scalability and also to dynamically adapt to the rapidly evolving (social) network scope.

Part II, "Data Dissemination in VSNs," composed of three chapters, is devoted to analyzing how data are transmitted over VSNs and all the related aspects, such as social clustering mechanisms, data dissemination, and connectivity analysis.

Chapter 3 discusses the social aspects of vehicles and describes novel social clustering mechanisms. Although there is a lot of research dedicated to VSNs, this work is mostly focused on the data sharing applications for infotainment. This chapter presents a novel vehicle clustering protocol that exploits the macroscopic social behavior of vehicles in order to create stable clusters for urban and highway scenarios. In Chapter 4, a novel social connectivity model for VSNs is presented. The social metrics of the communicating vehicles are then reviewed using the concepts of social theory, along with the conventional connectivity metrics in vehicular networks. Next, Chapter 5 presents a socially inspired broadcast data dissemination for vehicular ad hoc networks that exploits social features to determine when and which vehicles should rebroadcast data messages.

Part III, "Applications and Security Aspects in VSNs," composed of four chapters, investigates the main solutions for applications in VSNs, covering the topics of security and privacy, and data management strategies.

Chapter 6 introduces the concept of structural transitivity, that is, wireless social ties that connect one vehicle to other vehicles. The authors study the communications and carbon emission behaviors of vehicular networks. Simulation studies have confirmed that the wireless transitive relation among vehicles has a significant impact on the quality of service of wireless communications, as well as the carbon emissions generated by the vehicles. Moreover, this study reveals the correlations among topological transitivity, communication distance, and carbon emissions.

Chapter 7 presents vehicular crowdsourcing applications based on the distinct characterizations of VSNs. Specifically, this chapter briefly overviews VSNs and the challenges that affect the quality of vehicular crowdsourcing applications.

Chapter 8 develops an analytical model for assessing the probability of obtaining a suitable ride. The authors consider a generalization of the concept of ridesharing to tap into the underutilized transportation resources of the community. The chapter describes how a public transport service based on the transport commons concept may be implemented.

Finally, Chapter 9 deals with security and privacy requirements for VSNs. Awareness of those requirements is important in the context of VSNs so that strong security and privacy protection for the overall system can be ensured while deploying VSN applications. Moreover, a survey on the existing security and privacy solutions for emerging VSN applications is provided.

Target Audience

The main goal of this book is to target different people either interested in or working in the information and communication technology (ICT) domain. We have attempted to make it readable and designed it as a kind of guide in order to discover the several potentialities of the ICT and the social aspects in the context of the vehicular domain. In this sense, the targeted audience is not limited only to researchers and students, but also to include industry people from both the vehicular and ICT domains. The intention of the book is to show how integration of the social aspects in both domains, ICT and vehicular, can have a very high impact. The book provides significant technical and practical insights into different aspects, starting with a basic background on vehicular social architectures, going on to the data dissemination in VSNs and applications in the context of VSNs, and finally, discussing the future trends and technical challenges.

Acknowledgments

We want to thank the authors of the chapters of this book, who worked very hard to bring forward this unique resource on VSNs.

We would also like to thank the publishing and marketing staff of CRC Press/Taylor & Francis Group, in particular Richard O'Hanley, who worked with us on the project from the beginning. Finally, we would like to sincerely thank our respective families, for their continuous support and encouragement during the course of this project.

Editors

Valeria Loscrí has been a permanent researcher of the Future Ubiquitous Networks (FUN) Team in Inria Lille – Nord Europe since October 2013. She earned her master's degree in computer science and PhD in systems engineering and computer science in 2003 and 2007, respectively, both at the University of Calabria (Italy). In 2006, she spent 6 months as a visiting researcher at Rice University under the supervision of Professor Knightly, where she worked on the MAC layer of wireless mesh networks. She has authored more than 70 publications in journals, conferences, workshops and books. She is involved in several programs and organization committees such as Smart Wireless Access Networks for Smart cITY (SWANSITY), Body Area NanoNETworks: Electromagnetic, Materials and Communications (BANN-EMC), the IEEE International Conference on Wireless and Mobile Computing, Networking and Communications (WiMob), IDCS, International Conference on Computer Communication and Networks (ICCCN), and AdHocNow. Her research interests focus on performance evaluation, self-organizing systems, robotics networks, visible light communication, and nanocommunications. She is currently Scientific European Responsible for Inria Lille.

Athanasios V. Vasilakos is a professor at the Lulea University of Technology, Sweden. He served or is serving as an editor for many technical journals, including the *IEEE Transactions on Network and Service Management, IEEE Transactions on Cloud Computing, IEEE Transactions on Information Forensics and Security, IEEE Transactions on Cybernetics, IEEE Transactions on Nanobioscience, IEEE Transactions on Information Technology in Biomedicine, ACM Transactions on Autonomous and Adaptive Systems*, and *IEEE Journal on Selected Areas in Communications*. He is also general chair of the European Alliances for Innovation (www.eai.eu).

Anna Maria Vegni is non-tenured assistant professor in telecommunications at the Department of Engineering of Roma Tre University, Rome, Italy. She was awarded the laurea degree cum laude in electronics engineering in 2006 and earned her PhD degree in biomedical engineering, electromagnetic and telecommunications from Roma Tre University in 2010. In 2009, she was a visiting researcher in the Multimedia Communication Laboratory at the Department of Electrical and Computer Engineering, Boston University (Massachusetts, USA), under the supervision of Professor Little, where she worked on vehicular networking. Dr. Vegni is involved in several EU programs and organization committees such as the Visible Light Communications and Networking (VLCN) workshop at the 2015 IEEE International Conference on Communications and Body Area NanoNETworks: Electromagnetic, Materials and Communications (BANN-EMC) special track at 2015 International Conference on Body Area Networks (Bodynets). Her research activity focuses on vehicular networking, visible light communications, and nanocommunications. Since 2010, she has been in charge of the Telecommunications Networks Laboratory course at Roma Tre University.

Contributors

Syed Fakhar Abbas
School of Engineering, Computer and
    Mathematical Sciences
Auckland University of Technology
Auckland, New Zealand

Ozgur B. Akan
Koc University
Istanbul, Turkey

Adnan Al-Anbuky
School of Engineering, Computer and
    Mathematical Sciences
Auckland University of Technology
Auckland, New Zealand

Quan Bai
School of Engineering, Computer and
    Mathematical Sciences
Auckland University of Technology
Auckland, New Zealand

Pavlos Basaras
Department of Computer and
    Communication Engineering
University of Thessaly
Volos, Greece

Elif Bozkaya
Department of Computer Engineering
Istanbul Technical University
Istanbul, Turkey

Berk Canberk
Department of Computer Engineering
Istanbul Technical University
Istanbul, Turkey

Kardelen Cepni
Koc University
Istanbul, Turkey

Felipe D. Cunha
Federal University of Minas Gerais
Minas Gerais, Brazil

Mahmoud Hashem Eiza
School of Physical Sciences and Computing
University of Central Lancashire
Preston, United Kingdom

Flavio Esposito
Computer Science Department
Saint Louis University
St. Louis, Missouri

Helge Janicke
School of Computer Science and Informatics
De Montfort University
Leicester, United Kingdom

Hongyu Jin
Networked Systems Security Group
KTH Royal Institute of Technology
Stockholm, Sweden

Dimitrios Katsaros
Department of Computer and
    Communication Engineering
University of Thessaly
Volos, Greece

Mohammad Khodaei
Networked Systems Security Group
KTH Royal Institute of Technology
Stockholm, Sweden

William Liu
School of Engineering, Computer and
 Mathematical Sciences
Auckland University of Technology
Auckland, New Zealand

Antonio A. F. Loureiro
Federal University of Minas Gerais
Minas Gerais, Brazil

Leandros A. Maglaras
School of Computer Science and
 Informatics
De Montfort University
Leicester, United Kingdom

Guilherme Maia
Federal University of Minas Gerais
Minas Gerais, Brazil

Raquel Mini
Pontifical University of Minas Gerias
Minas Gerais, Brazil

Mustafa Ozger
Koc University
Istanbul, Turkey

Panos Papadimitratos
Networked Systems Security Group
KTH Royal Institute of Technology
Stockholm, Sweden

Farzad Safaei
Faculty of Engineering and Information
 Sciences
University of Wollongong
Wollongong, Australia

Qi Shi
Department of Computer Science Liverpool
John Moores University
Liverpool, United Kingdom

Aminu Bello Usman
School of Engineering, Computer and
 Mathematical Sciences
Auckland University of Technology
Auckland, New Zealand

Aline Carneiro Viana
INRIA
Saclay, France

Leandro Villas
University of Campinas
São Paulo, Brazil

INTRODUCTION TO VSNs

Chapter 1

Social Clustering of Vehicles

Leandros A. Maglaras and Helge Janicke

School of Computer Science and Informatics, De Montfort University, Leicester, United Kingdom

Pavlos Basaras and Dimitrios Katsaros

Department of Computer and Communication Engineering, University of Thessaly, Volos, Greece

Contents

1.1 Introduction

The vision of the vehicular ad hoc network (VANET) [1] is now very close to becoming reality and entering our everyday lives. Vehicles will be equipped with onboard units (OBUs), that is, a communication device that allows short-ranged wireless transmissions and hence facilitates vehicle-to-vehicle (V2V) communication. Roadside units (RSUs) will further aid the vehicular environment by serving as gateways to the Internet or other networks [2], and support numerous applications.

Benefits born of the VANET are numerous, and find plausible applications in the generic environment of information exchange. Examples include messages regarding traffic or weather conditions, hazard areas or road conditions, that is, safety applications, or infotainment services [3], allowing users on board vehicles to receive information relevant to services available in certain areas, for example, live video streaming and file sharing. Nonetheless, benefits do not come free of hazard. For instance, flooding the network with all kinds of messages will most likely exhaust the wireless network's resources, for example, cause severe contention and collisions, seize the very existence of the VANET, and thus abolish its benefits. Such challenges are addressed throughout the literature of vehicular networks in various ways [4]. Based on the fact that cars will be a major element of the expanding Internet of Things (IoT), with one in five vehicles having some sort of wireless network connection by 2020, the development of novel clustering mechanisms that exploit the augmented information gathered throughout the system is emerging.

The Social Internet of Things (SIoT) concept [5] is a network of intelligent objects that have social interactions. The Social Internet of Vehicles (SIoV) [6] is an example of an SIoT where the objects are smart vehicles. The authors in [7] describe a vehicular social network (VSN) as social interactions among cars that communicate autonomously to look for services (automaker patches or updates) and exchange information relevant to traffic. When moving on the social aspect of vehicular communications, new parameters like frequency of interactions between entities, historical data of driver behavior, and driver habits must be taken into account when creating groups of entities involved (e.g., cars, passengers, drivers, road and users).

This chapter discusses social aspects of vehicles and describes novel social clustering mechanisms. Although there is a lot of research dedicated to VSNs this work is mostly focused on the data sharing applications for infotainment. This chapter presents a novel vehicle clustering protocol that exploits the macroscopic social behavior of vehicles in order to create stable clusters for urban and high scenarios.

In particular, in this chapter we pursue two main goals: to analyze the different types of clustering methods and to present novel social clustering mechanisms for vehicles. Although many clustering methods for vehicular networks exist, they suffer from one or more of the following shortcomings: they are not generic in order to be applied in different environments, they are unpractical and difficult to be used in real-time situations, they do not exploit the road network topology, and they do not use the historic data in order to create stable clusters.

1.1.1 Motivation

Node clustering is proven to be an effective method to provide better data aggregation and scalability for wireless ad hoc networks. Clustering of nodes is also used, in order to cope with the interference caused by flooding of messages, since when the network is clustered, mainly the cluster head (CH) participates in the routing algorithm, which greatly reduces the number of necessary broadcasts. In vehicular networks, the social behavior of vehicles, that is, their tendency to share the same or similar routes, can be used in order to stabilize the clusters and increase their lifetime. The social characteristics of the drivers can help in the creation of more stable and robust clustering formations. The fact that each driver has some preferred routes that he or she tends to follow, depending on the time period, can be the basis for creating social profiles for each vehicle or driver based on historic trajectories of vehicles gathered by RSUs located throughout the road network. Combined communication capabilities along with social behavior of the vehicles can facilitate safety [8], eco-routing [9, 10], and infotainment [11], as well as dynamic charging of electric vehicle [12, 13] applications.

In this chapter, we present the steps that need to be taken in order to create social clusters of vehicles. The first step is the division of the road network into subnetworks in order to investigate each area in isolation. The partition is based on the connectivity among the road segments. The second step is the collection of the trajectories that the vehicles followed in these areas in order to create the social profiles (third step) for each of them, by using semi-Markov models. These profiles represent the behavior of each vehicle or driver in each area for each time period. The time periods are predawn (up until 8:00 a.m.), morning rush hour (8:00 a.m. to 10:00 a.m.), late morning (10:00 a.m. to noon), early afternoon (noon to 4:00 p.m.), evening rush hour (4:00 p.m. to 7:00 p.m.), and night time (after 7:00 p.m.). The last step is the exploitation of this enriched situation awareness, the social profiles of vehicles, in order to perform social clustering. A novel social clustering method for vehicles that follows these steps is presented, and its performance for different simulation settings, for example, communication range and velocity, is evaluated. The robustness of the method when some malicious vehicles tend to send bogus information due to infection is also investigated, and several defense mechanisms are proposed. The proposed social clustering of vehicles comes with low communication overhead and increased cluster stability.

1.2 Clustering of Vehicles and Vehicular Social Networks

Given the introduction of clustering in vehicular networks and its importance, this section briefly describes clustering techniques for vehicles [14] and the social aspect of vehicular communications [15, 16]. Note that we only discuss algorithms designed specifically for VANETS or techniques later improved to fit in the vehicular domain.

1.2.1 Clustering Vehicular Networks

In [17], the authors proposed a clustering technique based on a new aggregated local mobility (ALM) algorithm, with the objective to increase the stability of the devised clusters. The ratio of the received signal strength between two successive hello message exchanges, is used as a means to measure the relative variance in the mobility of the receiver's vicinity. When clustering (re)configuration stimulates, the vehicle node with the lower ALM, that is, the most stable node, is elected as CH. Finally, to avoid frequent (or unnecessary) cluster reorganization, additional time (and message exchanges) is needed in order to initiate reclustering. By minimizing the relative mobility and distance between CHs and their corresponding CMs, the algorithm was found to create stable clusters with a low average CH change rate.

In [18], enhanced spring clustering (ES-Cl) was proposed for highway environments. ES-Cl, apart from favoring vehicles with relative constant velocity or vehicles with predefined routes for CH selection, further incorporates a vehicle's physical dimensions, for example, its height. The intuition lies in the fact that communication can be much wider (less obstructed) when, for example, a tall vehicle acts as the transmitter. The results indicate that significant benefits can be obtained when the height of vehicles is taken into consideration for the CH selection process.

The authors in [19] utilize the affinity propagation algorithm [20] and propose a mobility-based clustering technique for VANETs. Their objective lies with low CH change rate and long duration of both CH and cluster member (CM) vehicle nodes. The algorithm is carried out with the use of paired exchanged messages, that is, responsibility and availability. These messages indicate the amount of degree to which a node j is exemplar to become a CH for another node i

(responsibility) and, respectively, *j*'s reply for its desire to comply with *i*'s request (availability). Responsibility and availability scores for each node are carried out by the affinity propagation algorithm with respect to its immediate vicinity, as the proposed method unfolds in a distributed manner. The algorithm was shown to meet its goals, that is, stable clusters with long-lasting members and CHs.

The ASPIRE [21] clustering algorithm aims at the provision of high network connectivity, low CH change rate, and high cluster size. Link criticality between pairs of neighboring vehicles is calculated and aggregated to obtain the criticality for each individual vehicle [22]. By exchanging their accumulated scores, vehicle nodes with low criticality are elected as CHs within their neighborhood. Unlike [22], ASPIRE is implemented in a local distributed fashion. It is a combination of varying components such as mobile nodes (e.g., vehicles) and stationary infrastructure (e.g., a base station). The evaluation results showed that vehicles are able to make better clustering decisions, even when the wireless medium is lossy.

The hierarchical clustering algorithm (HCA) was introduced in [23]. In contrast to the algorithms discussed so far, HCA does not require the knowledge of the vehicle nodes' location. Hence, the algorithm's robustness is enhanced since it does not rely on localization systems. Nonetheless, this is also a limiting factor because of the lack of knowledge for the mobility patterns of the vehicle nodes. The hierarchical fashion of the algorithm is organized as follows: at the lowest hierarchy there are the slave nodes at most two hops away from a CH; cluster relay (CR) nodes that relay messages from the CH to the slaves, and vice versa; and finally, the CH nodes that manage and synchronize the shared channel access for all other nodes in the formed cluster. The evaluation indicated, that even though HCA was originally designed for mobile ad hoc networks (MANETs), the formed clusters are stable and robust to topology changes, although with a few redundant clusters.

AMACAD as introduced in [24], takes into account the destination of vehicles, including the current location, speed, and finally, the relative and final destination of vehicles. These parameters will form the clusters in a distributed manner. AMACAD tries to closely follow the mobility patterns of neighboring vehicle nodes, in order to prolong the cluster duration. The results illustrated, highlight the benefits of considering the current location, the speed, and the vehicle's destination in the CH selection process.

VWCA was introduced in [25] and is based on [26]. The article proposes a clustering technique, along with another method, for detecting abnormal vehicle behaviors. Vehicles are characterized by a distrust value, and if deemed malicious, they are isolated. The authors base their clustering decisions on a weighed framework that accounts for the number of vehicle nodes in the vicinity, the distrust parameter, and the direction and entropy of the participating vehicles. The evaluation was conducted in highway scenarios. The results show that the accounted parameters can increase the stability and connectivity of the vehicular network.

1.2.2 Social Vehicles

Through vehicular communication, we allow vehicles to communicate with other cars within their proximity. However, so far, we have only looked at communicating vehicles as hollow entities. Taking into consideration the driver (and potential passengers) within the mobile units unlocks yet another dimension of the VANET. With this view, a driver can share data with his or her neighbors, that is, drivers and passengers in neighboring vehicles, based for example, on common interests, laying out the foundation for what is known as a VSN. Evidently, human behavior, for example, route preferences or individual selfishness, can have a large impact on the vehicular network, for example, its connectivity. With these considerations, researchers have shifted their focus in social aspects in order to improve the vehicular domain in both safety applications and infotainment.

Related works specifically addressed to vehicular social clustering are limited. Hence, we look at the broader sense and include relevant approaches that solve problems addressed also in clustering schemes, for example, relay selection and routing of data packets. For more details on binding social and vehicular networks, readers are referred to [15] and references therein.

Due to their very nature, VANETs are characterized as opportunistic networks. In order to alleviate the broadcast storm problem, opportunistic relay nodes are selected with the aim to reduce redundant retransmissions, that is, similarly to the function of a typical CH. In [27], the authors propose a method that deals with the aforesaid problem, namely, Selective Reliable Broadcast (SRB) protocol. SRB belongs to the class of broadcast protocols, as well as cluster-based approaches. The road network is classified into sectors, and vehicles can estimate their position through GPS. Neighboring vehicles are positioned in the same cluster if the distance between them is lower than a predefined threshold, and the farthest vehicle—with respect to the source—in each cluster is elected as the CH. The results illustrate that SRB works efficiently by means of a faster detection in congested areas.

BEEINFO was proposed in [28]. The authors take into consideration the social properties and mobility patterns of individuals and vehicles, in order to provide a socially aware approach for routing and forwarding data packets. The protocol considers a realistic environment composed of pedestrians, regular vehicles, and buses. The participants are classified in categories based on common interests, that is, communities, and regularities from mobility patterns, for example, the route of a bus, are exploited in order to choose proper relay nodes and disseminate information in a wider area. BEEINFO outperformed its competitors in terms of higher delivery ratio, less overhead, and less hop counts. The major drawback of the proposed method, as noted, is in some cases its average latency. Similar approaches that exploit social hot spots for selecting relay nodes while preserving human privacy include those reported in [29–31].

In [32], the authors note that protocols running on VANETs should be designed considering regular routines and mobility patterns, followed by drivers, as the day unfolds. The proposed algorithm chooses nodes to play the role of a relay based on tools used in social network analysis, that is, node degree and clustering coefficient. The results showed substantial performance gains when evaluating dense scenarios. Similar approaches exploiting centrality measures for selecting appropriate relay nodes include those in [33,34].

Overall, given the brief introduction of the socializing of vehicles concept, we understand that the incorporation of social driven aspects in the vehicular domain is an imminent evolution, which will change the way we will drive our cars in the years to come. In the sequence, we present our clustering algorithm, which incorporates characteristics of social behaviors, namely, sociological pattern clustering (SPC).

1.3 Social Clustering of Vehicles

In order to try to incorporate the social behavior of drivers in the clustering procedure, new parameters like rate of interactions among entities, historical data of driver behavior, and driver habits can be used. Based on the fact that people usually follow similar routes when moving in a city or on a highway for example, for commuting to work or driving children to school, we can build several social profiles of the drivers, one for every distinct social behavior. Using these social profiles, we can create stable clusters of vehicles that tend to follow similar routes.

Using the social behavior of vehicles as a basic parameter in the clustering formation procedure, we develop social clustering methods [35] that manage to increase cluster stability. The social behavior of drivers is derived from historical data that are collected from RSUs that are deployed at critical points of the road network. To implement the proposed clustering method, the path that vehicles are likely to follow is added to every beacon message. Social clustering uses the routes that the vehicles follow or tend to follow as an additional parameter to form stable clusters. In this chapter, we present the architecture of the SPC method, along with the main procedures that are involved.

1.3.1 Definition of the System

Definition 1.1 *Let $S = \{S_1, S_2, \ldots, S_M\}$ represent the set of road segments in a given geographical area or on a map, where each S_i is represented by a unidirectional edge between two consecutive junctions.*

Definition 1.2 *Let $V = \{V_1, V_2, \ldots, V_N\}$ be the set of vehicles that are traveling in the given geographical area during a certain time period.*

Definition 1.3 *Let $TP = \{TP_1, TP_2, \ldots, TP_K\}$ be the set of time periods that the investigated system is segmented into.*

1.3.2 Subnetworks and the Roles of RSUs

Road networks can be classified as complex networks, where the road segments and the junction connections do not follow regular patterns and may vary according to the geographical area. Identifying the communities that exist in a road network can help in dividing it into smaller subnetworks that can be loosely connected by intergroup edges. The subnetworks offer a better ground for the analysis and evaluation of novel applications.

An efficient partitioning technique was presented in [36], where the city is partitioned into small isolated parts. Based on this method, we cut a city in small subnetworks that can be investigated in isolation. In order to create the social profiles of the drivers and based on those profiles perform the social clustering, we assume that RSUs are deployed at every entry and exit point of each subnetwork. The RSUs are mainly used as data collection points, where approaching vehicles send their driving history through dedicated messages. These short messages (decentralized environmental notification message [DENM]) contain the set of road segments, also known as partial path PP_i that the vehicle V_i has traversed inside the subnetwork, along with the travel time T_j for each road segment, also known as partial time path PT_i. The format of each DENM message is shown in Figure 1.1. This information is used in order to create the transition table of each vehicle and, from this, to extract its social patterns.

Except for collecting useful information from vehicles that try to leave the subnetwork, RSUs also act as information points, where incoming vehicles are assigned a social number (SN) which is essential for the SPC method. This SN represents the social pattern that the vehicle is more likely

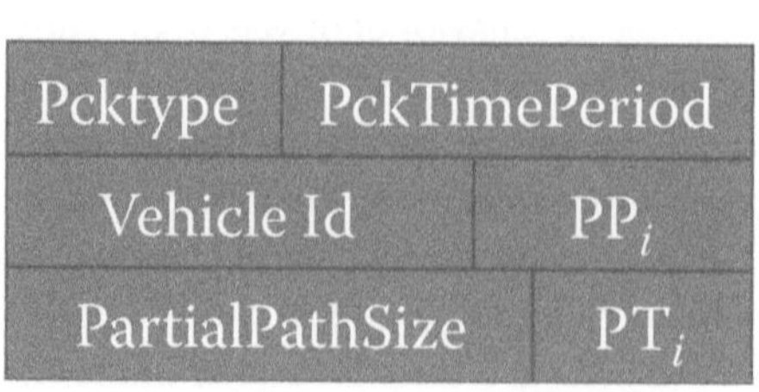

Pcktype	PckTimePeriod
Vehicle Id	PP_i
PartialPathSize	PT_i

Figure 1.1 Vehicle packet design. PP$_i$, partial path; PT$_i$, partial time path.

to follow based on the analysis of the historical data that the system has already collected for the specific vehicle. On the occasion that the vehicle is entering the subnetwork for the first time, it is assigned an SN that is the mean value for the specific time period.

1.3.3 Sociological Patterns

Every vehicle V_i, before leaving the subnetwork, sends to the nearest RSU all the info regarding the road segments that it traversed, along with the time. This information is inserted in a table PPT_{ik} that contains all the paths of the vehicle $V - i$ for the period TP_k. The mobility of the vehicle inside the subnetwork is represented as a semi-Markov process (Figure 1.2) and the transition probabilities from state to another are calculated using the information in table PPT_{ik}.

Based on these transition probabilities, we can extract the social pattern for the vehicle V_i for the specific time period TP_k (Figure 1.3).

Using this procedure, we create one social pattern for the vehicle V_i that is related to the specific time period TP and the road segment that it used in order to enter the subnetwork. Each social patterns is matched to a distinct SN. The outcome of this procedure is to create several social patterns for each vehicle that correspond to different driver behaviors. These different driver behaviors relate to the time of the day, such as driving to work in the morning and hobbies in the evening, and also the entry point in the subnetwork, which probably means a different final location. As also mentioned in the previous subsection, the next time that a vehicle enters the subnetwork, it is assigned an SN that matches better the entry point and current time. This number, as shown in the Section 1.3.4 is incorporated into the clustering mechanism in order to create clusters of vehicles that tend to follow the same routes.

$$\Lambda = \begin{pmatrix} 0 & a_{12} & 0 & a_{14} & 0 & 0 & 0 \\ 0 & 0 & a_{23} & a_{24} & 0 & 0 & 0 \\ 0 & 0 & 0 & 0 & 0 & a_{36} & a_{37} \\ 0 & 0 & a_{43} & 0 & a_{45} & a_{46} & 0 \\ 0 & 0 & 0 & 0 & 0 & a_{56} & 0 \\ 0 & 0 & 0 & 0 & 0 & 0 & a_{67} \\ 1 & 0 & 0 & 0 & 0 & 0 & 0 \end{pmatrix}$$

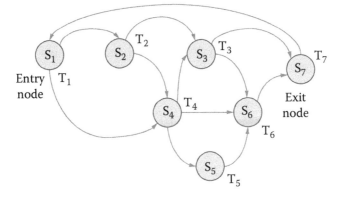

Figure 1.2 **Semi-Markov model of vehicle V_i in a simple network topology with one entry and one exit.**

$$A = \begin{pmatrix} 0 & a_{12} & 0 & a_{14} & 0 & 0 & 0 \\ 0 & 0 & a_{23} & a_{24} & 0 & 0 & 0 \\ 0 & 0 & 0 & 0 & 0 & a_{36} & a_{37} \\ 0 & 0 & a_{43} & 0 & a_{45} & a_{46} & 0 \\ 0 & 0 & 0 & 0 & 0 & a_{56} & 0 \\ 0 & 0 & 0 & 0 & 0 & 0 & a_{67} \\ 1 & 0 & 0 & 0 & 0 & 0 & 0 \end{pmatrix} \rightarrow \text{SP: } S_1, S_2, S_3, S_6, S_7$$

Figure 1.3 Transition table → social pattern.

1.3.4 Sociological Pattern Clustering Method

The SPC method uses the same basic clustering procedure as the virtual forces vehicular clustering (VFVC) method [18, 37, 38]. The vehicles are assigned virtual forces that are positive or negative according to the relative current and future positions that they have. Vehicles that are moving in the same direction or toward each other apply positive forces, while those traveling away from each other apply negative ones, and in order to perform clustering, the nodes periodically broadcast beacon messages. These cooperative awareness messages (CAMs) are used in order to inform surrounding vehicles about the host vehicles presence, and each consists of a: node identifier (*Vid*), node location, speed vector, total force *F*, status and time stamp. Each node *i* using the information of the beacon messages, calculates the pairwise relative force F_{relij} for every neighbor applied to every axis using the Coulomb law, according to Equation 1.1:

$$F_{relijx} = k_{ijx} \frac{q_i q_j}{r_{ij}^2}, \quad F_{relijy} = k_{ijy} \frac{q_i q_j}{r_{ij}^2} \tag{1.1}$$

where r_{ij} is the current distance among the nodes, and $k_{ijx}(k_{ijy})$ is a parameter indicating whether the force among the nodes is positive or negative, which depends on whether the vehicles are approaching or moving away from the corresponding axis. Parameters q_i and q_j could represent a special role for a node [37]. Every node computes the accumulated relative force applied to it along the axes *x* and *y* and the total magnitude of force *F*. According to the current state of the node and the relation of its *F* to its neighbor's, every node makes decisions about clustering formation, cluster maintenance, and role assignment.

On top of this mechanism, the *SN* of each vehicle is used in order to favor vehicles that share common paths in the subnetwork to create clusters. In order to do so, the SN of each vehicle is communicated with its neighbors, and during the clustering formation, it is used in order to split vehicles into clusters that have a common SN. Although this mechanism clearly increases the total number of clusters created, it is proven to be superior compared with other methods in terms of cluster lifetime and cluster stability.

The clustering formation procedure of the SPC method consists of two stages, where in the first stage, only vehicles that share common SN try to create clusters. In order to help isolated vehicles, in terms of *SN*, to join the created clusters there exists a second stage. During this stage, all vehicles participate in the clustering procedure regardless of their social pattern. Based on this clustering creation process, the different stages that a vehicle may have during its travel inside the subnetwork are represented in Figure 1.4.

In Figure 1.4, NS_i represents the set of all neighboring vehicles that share a common social pattern with vehicle V_i, while *UN* represents the undefined state that every vehicle has when initially joining or finally leaving the subnetwork. *Free* is the state that vehicle has when it has not joined a

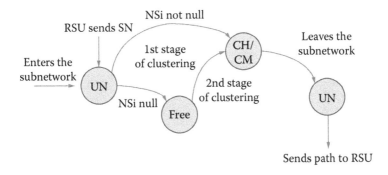

Figure 1.4   SPC clustering formation process.

cluster yet. *CH* represents the state of a vehicle that is elected as the CH in its neighborhood, and *CM* represents the CM state.

After the initial formation of the clusters, the SPC method has the clustering maintenance procedure that runs on each vehicle and tries to maintain the clustering structure of the network. During this procedure CMs that become free nodes, due to losing communication with their CH, try to join surrounding clusters. Also, when two or more CHs come within each other's transmission range and stay connected over a time threshold, their merge clusters into a bigger one, based on some rules.

1.4   Simulation and Performance Evaluation

In order to evaluate the performance of the SPC method, we simulated different mobility scenarios in the city of Erlangen, after having split it into subnetworks (Figure 1.5). Vehicles are assumed to communicate directly using wireless dedicated short-range communication (DSRC) channels. The power of the transmitter antenna is P_{tx} = 18 dBm and the communication frequency f is 5.9 GHz. In our simulations, we use a minimum sensitivity (*P*th) of 69–85 dBm, which gives a transmission range of 130–300 m. Simulation of vehicle mobility is conducted using simulation of urban mobility (SUMO) [39], and the produced traces are fed into our custom network simulator, where the different clustering methods are tested. In order to evaluate the performance of the SPC method we have implemented three established clustering methods, lowest ID (low-ID), dynamic Doppler value clustering (DDVC), and mobility prediction-based clustering (MPBC), which were proposed in [40], [41], and [42], respectively, All the simulation parameters with their default values are represented in Table 1.1.

VANETs are highly dynamic networks, and vehicles keep joining and leaving clusters as they travel along the road network. A good clustering method should ensure that most of the clusters that are created are stable. Cluster lifetime is an important metric for evaluating the performance of a clustering method and is directly related to the lifetime of the CH. The lifetime of a CH is defined as the time period from the moment that it is elected as a CH to the moment that it gives up its role.

1.4.1   Cluster Stability versus Communication Range

Figure 1.6 shows that mean cluster lifetime of the proposed method increases as the transmission range increases. A bigger communication range increases the probability of finding vehicles with a

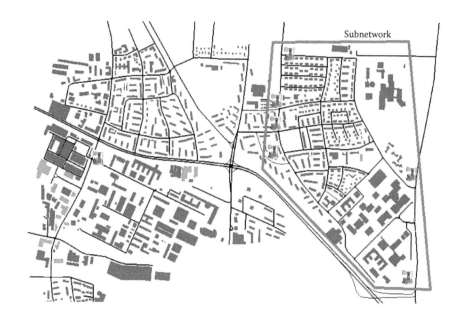

Figure 1.5   City is divided into subnetworks. RSUs are located at the entrances and exits of the subnetwork under investigation.

Table 1.1   Simulation Parameters

Independent Parameter	Range of Values	Default Value
Velocity (m/s)	20, 50	42
Number of vehicles	80, 120, 160	120
Communication range (m)	130–300	130
Number of RSUs	6	6

common SN in the vicinity, thus creating clusters through the first stage of the clustering procedure. Moreover, since SPC favors the creation of clusters of nodes that share common social profiles, as the mean communication range increases, the probability that such nodes stay interconnected for a longer time also increases.

For the low-ID method, where the vehicle's ID is the only parameter that is taken into account for electing the CH, an increased communication range does not improve its performance. This is due to the fact that although nodes' connectivity may be improved, the probability of meeting a neighbor with a lower ID and performing reclustering is also increased. The DDVC method, which takes into account the relative velocity among vehicles in order to spot the most central nodes and elect them as CHs, also stays unaffected from the increase in the communication range. This is happening since in urban environments, vehicles tend to frequently change directions and turn to different road segments, causing the method to perform reclustering. MPBC, which manages to perform the best among the competing methods, since it is designed to group nodes that move in random and independent directions, is still inferior compared with SPC, which creates more stable clusters by incorporating the social profile of drivers into the clustering mechanism.

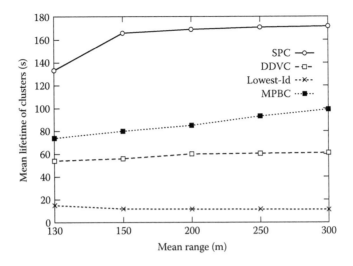

Figure 1.6 Lifetime vs. range.

1.4.2 *Cluster Stability versus Speed*

In order to evaluate the impact that different vehicle speeds have on the competing clustering methods, we performed another set of simulations. As shown in Figure 1.7, the impact of different speeds when simulating an urban environment is not so vivid. This is due to the fact that since most of the road segments are relatively small in length and vehicles have to constantly stop at intersections or lower their speed due to congestion, maximum velocities cannot easily be reached. So although the maximum velocity of simulated vehicles increased from 20 to 50 km/h, this did not heavily affect the stability of clusters. For all the simulations conducted SPC has much better performance than MPBC, DDVC, and low-ID. As discussed earlier, the improved performance of the proposed method is due to the incorporation of the social behavior of drivers, using the

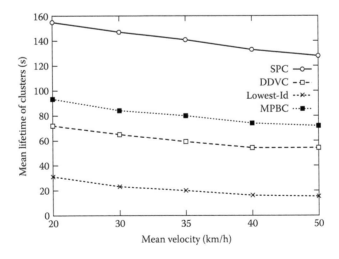

Figure 1.7 Lifetime vs. speed.

parameter SN, in the clustering formation procedure, thus leading to the creation of more stable clusters. The combination of the macroscopic information of the drivers (their habits) with the microscopic information (current and future direction of each vehicle) helps in the creation of more stable clusters, thus facilitating the communication among vehicles.

1.5 Social Vehicle Clustering under Attack Scenarios: Countermeasures

The area of security threats for VANETs has been widely investigated. Different attacks are possible such as denial-of-service (DoS) attack, fabrication attack, alteration attack, replay attack, message suppression attack, replay attack, and Sybil [43] attack. Different attackers exist in VANETs such as a selfish driver who tries to take advantage of the received information for personal benefit and a malicious attacker [44] that aims to harm the users or the network. DoS and distributed DoS attacks can affect the availability of the data, since the attacker can jam the medium, thereby disrupting the communication among the nodes [45]. It is obvious that such an attack would have an immediate effect on all the clustering methods, since they all rely on the exchange of messages among neighbors. The authors in [46] present a multiple-input multiple-output (MIMO)-based antijamming technique that uses rotation of the transmitted signal based on the instant jamming signal direction in order to restore the communication in the jammed area.

An attack that can have a major effect especially on the proposed SPC method is the false message infection, where an insider can send false information inside the network. Reports [*],[†] indicate that the infection of vehicles is now, indeed, a realistic scenario and the involvement of such in VANET protocols can result in catastrophic events. Since clustering algorithms are a basis for the development of routing methods [47, 48], tweaked data can strongly affect the performance of these methods in terms of delay and throughput.

On the occasion that tweaked information exists among the received data, performance of the clustering technique can degrade (Figure 1.8). In the simulated scenarios, the infected vehicles send wrong information regarding only their SN, causing the SPC method to create unstable clusters. As the number of infected vehicles increases, the performance of the method worsens. The rest of the competing clustering methods are not affected from this situation, since only fake SN scenarios were investigated. On the occasion where vehicles were disseminating fake information regarding their location or speed, then all clustering methods would be heavily influenced and the stability of the clusters would fall.

In order to suppress a false message infection attack, different defense mechanisms can be used. A first step toward devising an appropriate defense system is the ability to detect infiltrated vehicles. As noted in [49], misbehavior detection in VANETs can be divided into *node-centric* or *data-centric* mechanisms, with the first inspecting the behavior of a vehicle node, but not the data it sends. Other mechanisms in the same category include some form of reputation management, which inspects the past and present behavior of a node to derive the probability of future misbehavior, as implemented in [50]. Filtering out false data is another technique widely used in wireless sensor networks (WSNs) and VANETs [51]. Another recently introduced defense method is based on a combination of reputation and filtering [52]. All of these techniques can be incorporated in the SPC method in order to make it robust to false message infection.

---

* http://www.detroitnews.com/story/business/autos/2015/02/08/report-cars-vulnerable-wireless-hacking/23094215/
† http://www.techhive.com/article/221873/With_Hacking_Music_Can_Take_Control_of_Your_Car.html

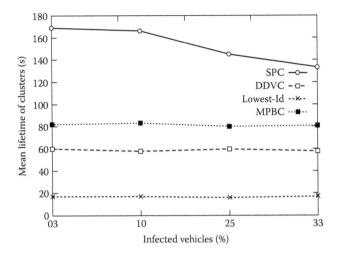

Figure 1.8 **False information affects SPC performance.**

1.6 Conclusions

In this chapter, we analyze different methods of clustering of vehicles that already exist and discuss the social aspects of vehicular communications. Based on this analysis, we present the steps that need to be taken in order to create social clusters of vehicles. The first step is the division of the road network into subnetworks in order to investigate each area in isolation. The partition is based on the connectivity among the road segments. The second step is the collection of the trajectories that the vehicles followed in these areas in order to create the social profiles (third step) for each one of them, by using semi-Markov models. These profiles represent the behavior of each vehicle or driver in each area for each time period.

We present a new social clustering method (SPC) for vehicles that incorporates these steps in order to create stable clusters, and we evaluate its performance for different simulation settings, for example, communication range and velocity. The obtained simulation results have demonstrated the greater effectiveness of SPC when compared with its competitors in terms of cluster stability. We finally investigate how robust our method is, when some malicious vehicles tend to send bogus information due to infection, and we discuss possible defense mechanisms that can suppress these attacks.

1.7 Glossary

CH: cluster head

CM: cluster member

DDVC: dynamic Doppler value clustering

DENM: decentralized environmental notification message

DoS: denial of service

Low-ID: lowest ID

MPBC: mobility prediction-based clustering

NS$_i$: neighborhood of vehicle i that share common social profile

$\mathbf{P}_{tx}$: power of the transmitter antenna

$\mathbf{PP}_i$: partial path

$\mathbf{PPT}_{ik}$: path table of vehicle i for time period k

$\mathbf{PT}_i$: partial time path

$\mathbf{P}_{th}$: minimum sensitivity

RSU: road side unit

SN: social number

SPC: sociological pattern clustering

$\mathbf{TP}_K$: time period K

VANET: vehicular ad hoc network

VFVC: virtual forces vehicular clustering

UN: undefined state

References

1. S. Zeadally, T. Hunt, Y. S. Chen, A. Irwin, and A. Hassan. Vehicular ad hoc networks (vanets): Status, results, and challenges. *Telecommunication Systems*, 50(4):217–241, 2010.

2. Z. H. Mir and F. Filali. LTE and IEEE 802.11p for vehicular networking: A performance evaluation. *EURASIP Journal on Wireless Communications and Networking*, 2014(1):1–15, 2014.

3. H. T. Cheng, S. Hangguan, and S. Weihua. Infotainment and road safety service support in vehicular networking: From a communication perspective. *Mechanical Systems and Signal Processing*, 25(6):2020–2038, 2011.

4. R. Naja, ed *Wireless Vehicular Networks for Car Collision Avoidance*. Vol. 2013. Berlin: Springer, 2013.

5. L. Atzori, A. Iera, G. Morabito, and M. Nitti. The social Internet of things (siot)–When social networks meet the Internet of things: Concept, architecture and network characterization. *Computer Networks*, 56(16):3594–3608, 2012.

6. K. M. Alam, M. Saini, and A. El Saddik. Towards social Internet of vehicles: Concept, architecture and applications.

7. M. Nitti, R. Girau, A. Floris, and L. Atzori. On adding the social dimension to the Internet of vehicles: Friendship and middleware. In *Communications and Networking (BlackSeaCom), 2014 IEEE International Black Sea Conference on*, pp. 134–138. IEEE, 2014.

8. J. Barrachina, P. Garrido, M. Fogue, F. J. Martinez, J.-C. Cano, C. T. Calafate, and P. Manzoni. Reducing emergency services arrival time by using vehicular communications and evolution strategies. *Expert Systems with Applications*, 41(4):1206–1217, 2014.

9. K. Boriboonsomsin, M. J. Barth, W. Zhu, and A. Vu. Eco-routing navigation system based on multi-source historical and real-time traffic information. *IEEE Transactions on Intelligent Transportation Systems*, , 13(4):1694–1704, 2012.

10. L. Zi-fa, Z. Wei, J. Xing, and L. Ke. Optimal planning of charging station for electric vehicle based on particle swarm optimization. In *Innovative Smart Grid Technologies–Asia (ISGT Asia)*, pp. 1–5. IEEE, 2012.

11. S. Smaldone, L. Han, P. Shankar, and L. Iftode. Roadspeak: enabling voice chat on roadways using vehicular social networks. In *Proceedings of the 1st Workshop on Social Network Systems*, pp. 43–48. New York: ACM, 2008.

12. L. Maglaras, F. V. Topalis, A. L. Maglaras. Cooperative approaches for dymanic wireless charging of electric vehicles in a smart city. In *2014 IEEE International Energy Conference (ENERGYCON)*, pp. 1365–1369. IEEE, 2014.

13. P. Dutta. Coordinating rendezvous points for inductive power transfer between electric vehicles to increase effective driving distance. In *2013 International Conference on Connected Vehicles and Expo (ICCVE)*, pp. 649–653. Piscataway, NJ: IEEE, 2013.

14. S. Vodopivec, J. Bešter, and A. Kos. A survey on clustering algorithms for vehicular ad-hoc networks. In *2012 35th International Conference on Telecommunications and Signal Processing (TSP)*, pp. 52–56. Piscataway, NJ: IEEE, 2012.

15. A. M. Vegni and V. Loscri. A survey on vehicular social networks. IEEE Communications Surveys & Tutorials, 17(4):2397–2419, 2015.

16. L. A. Maglaras, A. H. Al-Bayatti, Y. He, I. Wagner, and H. Janicke. Social Internet of vehicles for smart cities. *Journal of Sensor and Actuator Networks*, 5(1):3, 2016.

17. E. Souza, I. Nikolaidis, and P. Gburzynski. A new aggregate local mobility (ALM) clustering algorithm for VANETS. In *Communications (ICC), 2010 IEEE International Conference on*, pp. 1–5. Piscataway, NJ: IEEE, May 2010.

18. L. A. Maglaras and D. Katsaros. Enhanced spring clustering in VANETs with obstruction considerations. In *Proceedings of the IEEE Vehicular Technology Conference—Spring (VTC Spring)*, 2013.

19. C. Shea, B. Hassanabadi, and S. Valaee. Mobility-based clustering in VANETs using affinity propagation. In *IEEE Global Telecommunications Conference (GLOBECOM)*, pp. 1–6, 2009.

20. J. F. Brendan and D. Dueck. Clustering by passing messages between data points. *Science*, 315:972–976, 207.

21. A. Koulakezian. ASPIRE: Adaptive service provider infrastructure for VANETs. Master thesis, University of Toronto, 2011.

22. A. Tizghadam and A. Leon-Garcia. Survival value of communication networks. In *IEEE INFOCOM Workshops 2009*, pp. 1–6. Piscataway, NJ: IEEE, April 2009.

23. E. Dror, C. Avin, and Z. Lotker. Fast randomized algorithm for hierarchical clustering in vehicular ad-hoc networks. In *10th IFIP Annual Mediterranean Ad Hoc Networking Workshop (Med-Hoc-Net)*, pp. 1–8. Piscataway, NJ: IEEE, 2011.

24. M. M. C. Morales, S. H. Choong, and C. B. Young. An adaptable mobility-aware clustering algorithm in vehicular networks. In *13th Asia-Pacific Network Operations and Management Symposium (APNOMS)*, pp. 1–6. Piscataway, NJ: IEEE, September 2011.

25. D. Ameneh, G. P. R. Akbar, and K. Ahmad. VWCA: An efficient clustering algorithm in vehicular ad hoc networks. *Journal of Network and Computer Applications*, 34(1):207–222, 2011.

26. M. Chatterjee, S. K. Das, and D. Turgut. WCA: A weighted clustering algorithm for mobile ad hoc networks. *Cluster Computing*, 5(2):193–204, 2002.

27. A. M. Vegni and E. Natalizio. Forwarder smart selection protocol for limitation of broadcast storm problem. *Journal of Network and Computer Applications*, 47:61–71, 2015.

28. X. Feng, L. Li, L. Jie, A.M. Ahmed, L.T. Yang, and Jianhua M. Beeinfo: Interest-based forwarding using artificial bee colony for socially aware networking. *IEEE Transactions on Vehicular Technology*, 64(3):1188–1200, 2015.

29. L. Rongxing, L. Xiaodong, L. Xiaohui, and S. Xuemin. Sacrificing the plum tree for the peach tree: A socialspot tactic for protecting receiver-location privacy in VANET. In *IEEE Global Telecommunications Conference (GLOBECOM)*, pp. 1–5. Piscataway, NJ: IEEE, 2010.

30. L. Rongxing, L. Xiaodong, T.H. Luan, L. Xiaohui, and S. Xuemin. Pseudonym changing at social spots: An effective strategy for location privacy in VANETs. *IEEE Transactions on Vehicular Technology*, 61(1):86–96, 2012.

31. L. Xiaodong, L. Rongxing, L. Xiaohui, and S. Xuemin. Stap: A social-tier-assisted packet forwarding protocol for achieving receiver-location privacy preservation in VANETs. In *2011 Proceedings of IEEE INFOCOM*, pp. 2147–2155. Piscataway, NJ: IEEE, 2011.

32. F. D. Cunha, G. G. Maia, A. C. Viana, R. A. Mini, L. A. Villas, and A. A. Loureiro. Socially inspired data dissemination for vehicular ad hoc networks. In *Proceedings of the 17th ACM International Conference on Modeling, Analysis and Simulation of Wireless and Mobile Systems (MSWiM '14)*, pp. 81–85. New York: ACM, 2014.

33. A. Stagkopoulou, P. Basaras, and D. Katsaros. A social-based approach for message dissemination in vehicular ad hoc networks. In N. Mitton, A. Gallais, M. E. Kantarci, and S. Papavassiliou, eds., *Ad Hoc Networks*. Vol. 140 of Lecture Notes of the Institute for Computer Sciences, Social Informatics and Telecommunications Engineering, pp. 27–38. Berlin: Springer , 2014.

34. A. Bradai and T. Ahmed. ReViV: Selective rebroadcast mechanism for video streaming over VANET. In *2014 IEEE 79th Vehicular Technology Conference (VTC Spring)*, pp. 1–6Piscataway, NJ: IEEE, 2014.

35. L. Maglaras and D. Katsaros. Social clustering of vehicles based on semi-Markov processes. *IEEE Transactions on Vehicular Technology*, 65(1):318–332, 2016.

36. Q. Song and X. Wang. Efficient routing on large road networks using hierarchical communities. *IEEE Transactions on Intelligent Transportation Systems*, 12(1):132–140, 2011.

37. L. A. Maglaras and D. Katsaros. Distributed clustering in vehicular networks. In *Proceedings of the IEEE International Conference on Wireless and Mobile Computing, Networking and Communications (WiMob)*, pp. 593–599. Piscataway, NJ: IEEE, 2012.

38. L. A. Maglaras and D. Katsaros. Clustering in urban environments: Virtual forces applied to vehicles. In *Proceedings of the IEEE Workshop on Emerging Vehicular Networks: V2V/V2I and Railroad Communications*. Piscataway, NJ: IEEE, 2013.

39. M. Behrisch, L. Bieker, J. Erdmann, and D. Krajzewicz. SUMO–Simulation of urban mobility. In *Third International Conference on Advances in System Simulation (SIMUL 2011)*. Wilmington, DE: IARIA, 2011.

40. M. Gerla and J. T. Tsai. Multicluster, mobile, multimedia radio network. *Wireless Networks*, 1(3):255–265, 1995.

41. E. Sakhaee and A. Jamalipour. Stable clustering and communications in pseudolinear highly mobile ad hoc networks. *IEEE Transactions on Vehicular Technology*, 57(6):3769–3777, 2008.

42. M. Ni, Z. Zhong, and D. Zhao. MPBC: A mobility prediction-based clustering scheme for ad hoc networks. *IEEE Transactions on Vehicular Technology*, 60(9):4549–4559, 2011.

43. J. R. Douceur. The sybil attack. In *Peer-to-Peer Systems*, pp. 251–260. Berlin: Springer, 2002.

44. L. Buttyán, T. Holczer, and I. Vajda. On the effectiveness of changing pseudonyms to provide location privacy in VANETs. In *Security and Privacy in Ad-Hoc and Sensor Networks*, pp. 129–141. Berlin: Springer, 2007.

45. O. Punal, C. Pereira, A. Aguiar, and J. Gross. Experimental characterization and modeling of RF jamming attacks on VANETs. *IEEE Transactions on Vehicular Technology*, 64(2):524–540, February 2015.

46. Q. Yan, H. Zeng, T. Jiang, M. Li, W. Lou, and Y. T. Hou. MIMO-based jamming resilient communication in wireless networks. In *2014 Proceedings of IEEE INFOCOM*, pp. 2697–2706. Piscataway, NJ: IEEE, 2014.

47. D. R. Girinath and S. Selvan. A novel cluster based routing algorithm for hybrid mobility model in VANET. *International Journal of Computer Applications*, 1(15):35–42, 2010.

48. Y. Luo, Wei Zhang, and Yangqing Hu. A new cluster based routing protocol for VANET. In *2010 Second International Conference on Networks Security, Wireless Communications and Trusted Computing (NSWCTC)*, vol. 1, pp. 176–180. Piscataway, NJ: IEEE, 2010.

49. U. Khan, S. Agrawal, and S. Silakari. A detailed survey on misbehavior node detection techniques in vehicular ad hoc networks. In *Information Systems Design and Intelligent Applications*, pp. 11–19. Berlin: Springer, 2015.

50. CH. Kim and IH. Bae. A misbehavior-based reputation management system for vanets. In *Embedded and Multimedia Computing Technology and Service*, pp. 441–450. Springer, 2012.

51. Z. Cao, J. Kong, U. Lee, M. Gerla, and Z. Chen. Proof-of-relevance: Filtering false data via authentic consensus in vehicle ad-hoc networks. In *INFOCOM Workshops 2008*, pp. 1–6. Piscataway, NJ: IEEE, 2008.

52. P. Basaras, L. Maglaras, D. Katsaros, and H. Janicke. A robust eco-routing protocol against malicious data in vehicular networks. In *2015 8th IFIP Wireless and Mobile Networking Conference (WMNC)*, pp. 184–191. Piscataway, NJ: IEEE, 2015.

Chapter 2

Vehicular Social Sensor Networks

Kardelen Cepni, Mustafa Ozger, and Ozgur B. Akan

Koc University, Istanbul, Turkey

Contents

2.1 Introduction

Online social networks (OSNs) have recently become the essential means of networking, communication, and entertainment. OSNs provide web-based services, in which users create their profiles and communicate with their friends in the network [1]. Many OSNs worldwide provide distinct functionalities and applications [2].

Frequent use of OSNs by mobile users leads to the integration of social features with wireless networks in the domain of vehicles. For example, drivers on the roads can form a social network to share information related to the road conditions and to communicate with each other. To this end, the vehicular social network (VSN) was first proposed in a seminal paper [3], in which it integrates the communication infrastructure and social needs of the people in the vehicles in such a way that they can be socialized. For example, mobile users may use an OSN service to share information with the other users in their network. While the main component of a VSN is the vehicle equipped with an advanced technological interface, the information sharing activities in the network are mostly managed by the users, who are connected based on a mutual interest, for example, friendship, location or coworker.

In this chapter, we introduce a novel paradigm, named vehicular social sensor networks (VSSNs), to explore social sensing mechanisms in VSNs and investigate their social sensing capabilities in VSNs. To this end, we first introduce the concept of social sensing and the clear definition of VSSNs. We also point out the challenges posed by VSSNs and possible application areas. Networking model, the key elements of VSSNs, and the social sensing mechanisms are explained in a detailed manner. Furthermore, we investigate the sensor-related factors in the reliability of VSSNs to extract the event information from the VSSN nodes. In the last section, we focus on the characteristics of sensory data produced by VSSN nodes and their utilization for social sensing.

The contribution of this chapter is mainly to introduce a new networking paradigm, which combines online social sensing, OSNs and vehicular networks. Toward this end, we explain social sensing via OSNs in vehicular networks and our proposed paradigm in detail in the next subsections.

2.1.1 Social Sensing

The emergence of OSNs and generation of information through users' mobile devices via OSNs have improved the applications of Big Data, which is a massive volume of either structured or unstructured data that comes from multiple sources such as documents, the public web, social media, or sensor data. Information generation through mobile devices resulted in a new approach named Mobile Big Data (MBD), which mostly encompasses the datasets generated by moving devices, for example, vehicles and mobile phones. People mostly use OSNs via their mobile devices for information retrieval and to share, for example, their observations or feelings about a subject, with their contacts and contribute to the generation of MBD. This behavior makes OSNs one of the biggest sources of MBD. It may include more variety in content and volume than the Big Data since information is mostly shared by the mobile users from different locations at different times.

On the other hand, a tremendous amount of MBD that originated from the frequent use of OSNs during major events led to a new application, known as *social sensing*, which is the utilization of the MBD, that is, information shared in OSNs, to estimate an observed yet unknown phenomenon. This application has fostered the research on both OSNs and MBD since social sensing has a big societal impact in many aspects of human lives. For instance, during social disasters the authorities may feel the need to investigate users' perceptions, that is, learn from collected observations, and they can design their emergency plan accordingly to maintain control. More examples can be given in other fields, including online marketing, politics, and financial markets. Therefore, understanding the characteristics of data shared in OSNs and the social sensing capabilities of OSNs becomes an important problem.

Social sensing involves an estimation problem, that is, given a collection of users' observations, that is, MBD set, an unknown parameter of interest is aimed to be estimated. Social sensing has been studied from several aspects. The first line considers social sensing an application-specific estimation problem and aims to develop event estimation and detection algorithms using the data in OSNs. For example, Twitter messages are used for a real-time heavy traffic detection application [4], and an earthquake detection system is proposed based on the sentiment analysis of Twitter messages [5]. The approach is likely to be beneficial in developing social sensing applications in real life; however, it does not reflect the real nature of social sensing since the algorithms used in social sensing may modify or eliminate the existing data after analysis, for example, sentiment analysis. If there is repetitive, inaccurate, or irrelevant information in the dataset, then the algorithms eliminate it to improve the estimation. The second line introduces social sensing as an abstract concept by defining the users as *social sensors* with unreliable sensing capabilities. Several studies build an introductory framework for social sensing and introduce its future applications [6–8].

However, there is still a lack of theoretical understanding of MBD and its relation to social sensing. To the best of our knowledge, none of these studies have explored the impact of users and also the information sharing processes in OSNs on MBD and social sensing. We study social sensing with Twitter from a communication theoretical perspective in [9]. By modeling Twitter as a sensor network and information transmission as an on–off channel with constant delay, we investigate the effect of the features of Twitter and channel characteristics on the reliability of social sensing. Our results indicate that different user behaviors and network characteristics, for example, delay levels and on–off characteristics, have an effect on the performance of social sensing.

MBD is mainly generated from the user interactions in VSNs, and the dataset is utilized for social sensing. From a sensor networking perspective, formation of MBD resembles wireless sensor networks (WSNs), in which users exchange or share information with their contacts and shared information in the OSN is used in estimation of a phenomenon. As an analogy, the users in OSNs can behave as sensors in WSNs. Whether a WSN is centralized or distributed, each sensor node shares its observation to a common sink. A sensor node in a WSN can be regarded as the user in OSN, and the sink in the WSN can be regarded as the server of the OSN. According to the collected observations, the sink in the WSN estimates the event, which is also the case in OSNs. VSN is also a social network; however, its domain is confined to vehicular networks. The relationship between OSNs and WSNs is the same as the relationship between VSNs and WSNs. Hence, the collection of information by the sensors due to monitoring applications resembles the collection of information from the VSN users to estimate a social event. In this type of VSN, the aim of the network is not socializing but extracting event information from the other VSN nodes. Features of OSNs define how the information is exchanged, and features of VSN define mobility characteristics of the data. Moreover, the characteristics of the observation signals may change depending on the source signal characteristics, user behaviors, and features of the OSN and VSN. Therefore, further studying MBD in VSNs enabled by OSNs is imperative in both characterization of the social sensing capabilities of VSNs and the development of efficient social sensing applications.

Besides formation of MBD for social sensing, mobile users can extract the information related to events. The extraction of information is possible by forming peer-to-peer networks among the social sensors. This type of network formation does not need the MDB approach; hence, each user can gather the information related to the events from its neighbors or friends, who are the social sensors sensing those social events. It supports distributed social sensing since the social sensor observations about an event are not collected in a centralized server. To completely define the social sensing mechanism within VSNs enabled by OSNs, we introduce the VSSN paradigm in Section 2.1.2.

2.1.2 Definition of Vehicular Social Sensor Networks

Social sensing is the estimation of events through online social networking. Hence, the sensing capabilities of OSN users have resulted in different networking applications. The application of sensor networking applications to mobile networks brings new challenges and opportunities. For instance, [10] discusses the integration of the sensor networking paradigm with vehicular networks. The authors in [10] explain that VSNs use the advantages of the mobility of vehicles and the sensing capabilities of sensor networks. Furthermore, it also proposes to bridge vehicular sensor networks with social networks. Vehicular sensor networks collect real-time data, and social networks provide the ability to share information among the users in different locations. This bridging enables the combination of data coming from the vehicular sensor networks and social networks. Application-specific information can also be from the collected data via estimation.

The relationship between WSNs and OSNs reveals that some of the functions of WSNs, such as event estimation, can be utilized for social events in OSNs. The combination of WSNs and OSNs may be used in different domains. One of the most important is the vehicular domain. We propose a new networking paradigm, which is VSSN, to support mobility and sensing for online social networking. VSNs use social networking to enable communication between users with mobility [11]. Hence, VSSNs can be regarded as VSNs with the capability of extracting event information by the collective efforts of VSN nodes. The users in the vehicles can socialize by communicating with their neighbor vehicles or distant vehicles in VSNs, and these users can gain information about events on the roads. Hence, we may define a VSSN as a *kind of VSN that consists of mobile social sensor nodes that are capable of sensing an event with their sensory mechanisms and reporting their observation signals through the information dissemination mechanisms in the VSN via OSN tools wirelessly.*

A VSSN consists of three elements: (1) the vehicular communication channel, (2) a mobile network consisting of social nodes that sense the environment and share their observations via the social networking service on the application layer, and (3) the estimation of the unknown phenomenon by using the available information in the VSSN. As VSNs contribute to the MBD set, which can be utilized for social sensing, VSSNs are also the key contributor to MBD. Hence, exploration of VSSNs is essential for understanding the reliability of event estimation.

A network architecture of a VSSN can be seen in Figure 2.1, which also shows the relationship between OSNs and VSNs. VSN is a subset of OSN since the users of a VSNs are limited to being on vehicles, while OSN users do not have such a limitation. We show some part of a VSN consisting of a few cars in Figure 2.1. These cars communicate with each other and base stations. The colored arrows indicate these communication links, which contain the information about the event they sense and observe. VSSN is the combination of the VSN and sensing mechanism as seen in Figure 2.1. There are two mechanisms for the social sensing. The first one is distributed sensing, where a VSSN user extracts the event information from the observation of its neighbors. The second one is to extract the event information from the collected data in MBD, which contains all the information related to the event from the users. The three elements of VSSN can be clearly seen in Figure 2.1, which are the channel, network, and sensing mechanism.

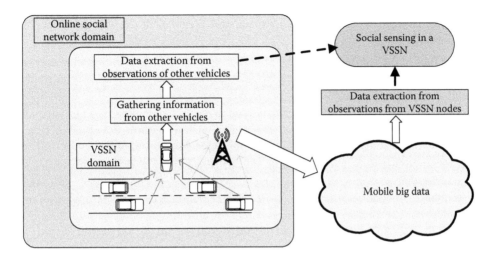

Figure 2.1 **VSSN architecture and its relation to OSNs and VSNs.**

2.1.3 Possible Application Areas

VSSNs provide dynamic social sensing mechanisms. Movement of the users causes changes in the number of social nodes that participate in social sensing frequently. Furthermore, the link conditions and neighboring relations change frequently due to the mobility of social sensors. VSSN nodes are also in direct interaction with other VSSN nodes and the environment. Hence, the dynamic network architecture and direct interaction with the environment result in diverse application areas. Possible applications of VSSNs are as follows:

- Up-to-date traffic information via vehicular social sensing and the determination of the traffic paths according to the traffic and road conditions
- Real-time weather condition tracking based on information exchange in the VSSNs formed within a local region
- Real-time spot (on roads) measurements via reporting of the social sensors in vehicles to avoid traffic accidents beforehand
- Real-time event detection via social sensors; for example, drivers in the opposite lane may inform each other that there is an incident (this can be regarded as an example of a distributed social sensing mechanism with peer-to-peer communications)
- Real-time information sharing via social sensors; for example, drivers may share the possible parking spots in a crowded area to help other drivers
- Peer-to-peer vehicular social sensing on the road in case of no Internet connection

2.1.4 Challenges Posed by VSSNs

The dynamic conditions of VSSNs due to the mobility and behaviors of the social sensors are the main challenges posed by VSSNs. These challenges are as follows:

- The reliability of the mobile sensor nodes in VSSN depends on the users since the mobile users behave as sensors by choosing what information and when to share in the VSSN.

The behavior of the users is very important since they provide the data to extract the event information.

■ The time of the sharing is of utmost important since the provided information may become obsolete for events having short a life span. Furthermore, the time delay in sharing the information introduces distortion in the collected data.

■ The observation data may not be reliable due to malicious mobile users. The false information provided by the mobile sensors may deteriorate the performance of the social sensing. Hence, eliminating these types of users poses a challenge in VSSNs.

■ The communication channel contains the challenges posed by the mobility of the users in vehicles. Access to the Internet in vehicles may have some options, such as through base stations or mobile relay stations. Sensor observations are gathered through this channel, which may experience random delays due to the channel.

■ Observation data generated in VSSNs (MBD) poses challenges for event estimation due to the volume and dynamic characteristics of the data and requires design of efficient aggregate data analysis and estimation algorithms in terms of scalability and speed.

2.2 Sensor Networking in Vehicular Social Sensor Networks

In this section, we investigate the sensor networking approach and key elements in VSSNs to realize the vehicular social sensor networking. The users in vehicles deliver their observations to the server or other users in neighboring vehicles through the wireless channel. From this point of view, the user in the vehicle can be considered an OSN user willing to share his or her observations. This can be modeled as in Figure 2.2.

2.2.1 Networking Model

An event on the road is sensed by the users in vehicles. The event is denoted by θ. This event is perceived by the user as S_k, where k represents the index of the social sensor. The perceived message is encoded depending on the application used by the user. The perception of the users about the event is not always objective. Hence, the perceived event may be distorted by the users, which degrades the quality of social sensing. The encoded message may be in the form of text, video, image, or a combination of them. The OSN application will form the packets accordingly.

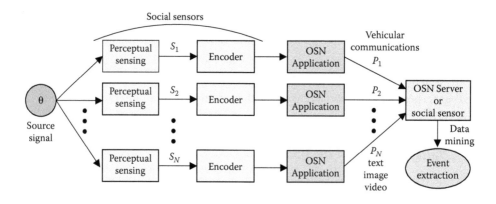

Figure 2.2   Networking model for VSSNs.

The encoded packet is delivered according to the communication types, which are vehicle-to-vehicle (V2V) and V2I communication. This actually represents the wireless channel between the communicating vehicles or between the vehicle and the base station. The encoded message may be distorted during the transmission. The Internet connection blackouts may be present, and delay may increase due to the mobility of the users. These packets are received by the OSN server or a social sensor to extract the event, as seen in Figure 2.2. In Section 2.2.2, we investigate source signal and social sensors in VSSNs, which are the elements in VSSNs.

2.2.2 Key Elements of VSSNs

2.2.2.1 Source Signal

The main motivation of social sensing is to estimate the source signal. The source signal is about the event of interest. The source event may be accidents on roads, the conditions of the road in specific areas, traffic jams or events not related to the vehicles and road conditions, such as political or sport events.

Modeling of the event is important to extract it from the observations. The event models may be in one of two forms: binary and continuous. If there is a traffic accident in a certain spot, this event is modeled as a binary signal. An event of traffic congestion is modeled as a continuous signal since it continues for some time.

The interaction between the vehicle users and the source signals reveals a new categorization of events. Static events are independent of the users. For example, foggy conditions or icy roads are examples of static events; the source signal is not affected by the users. On the other hand, dynamic events are those where the vehicle users can change the event characteristics. For instance, the source signal indicating that there is traffic congestion on a certain road affects the drivers' behaviors. The informed drivers do not use this road to avoid the congestion.

The extraction of the event from the user observations needs a complete characterization and modeling of the source signals. There exist open issues to address the modeling of the source signals, which are outlined as follows:

- A general framework should be established to characterize the source signal considering the on–off nature and continuity of the source signals.

- The effect of the source signal modeling should be thoroughly investigated to extract the information for the social sensing.

- The continuity and discreteness of the source signal should be investigated in regard to social sensing. According to this characterization, event estimation reliability levels may be studied.

2.2.2.2 Social Sensors

The information sources of the OSNs are the social sensors. These sensors are the users in the vehicles, and they sense the event and sample it accordingly. The perception of the event by a user in a vehicle may be through the sensing capabilities or the sensors deployed in the vehicle. Hence, the information is fed to the OSN server or to another neighboring vehicle by the users in the vehicles.

The user in VSSNs can observe the traffic congestion and, for instance, tweets about it using V2I communication. The users that extract this information can avoid the traffic jam. For instance, the user in some part of the road can inform the other users through V2V communication about the fact that an ambulance is approaching in order to decrease the time to reach the hospital.

These examples state that the users are important players in the social sensing in VSSNs. However, their role in social sensing may be adversarial due to misinterpretation and misperception. This type of behavior may deteriorate the accuracy of social sensing. This problem can be overcome by devising a reputation mechanism. This mechanism can rank the users according to their trustworthiness [12].

The vehicle may provide false reports about the conditions of the road due to the noise. It may also distort the message. This situation also degrades the quality of social sensing.

The format of the packet from the observed event depends on the preferences of OSN users and the OSN application. The source format of the information may be different depending on the applications used. For instance, Twitter supports the messages in text, picture, and video format. Hashtags, which are the keywords about the event, give detailed information about it. Facebook also gives the opportunity to share the messages by the users in the vehicles. Each form of the message has different types and levels of error. These factors are evaluated in Section 2.4, which explains the reliability of event estimation in VSSNs.

2.3 Social Sensing Schemes in VSSNs

Social sensing in vehicular networks enables valuable information about the conditions of the road and events on the roads to be provided to the users in vehicles. The sensing information can be spread by the social sensor in the vehicles via the social networks. Social sensing may be divided into two categories due to the differences in information extraction process. These social sensing schemes are centralized social sensing and distributed social sensing. In this section, we investigate the social sensing schemes in VSSNs from the perspective of the sensor networks by giving an overview of some studies in literature and proposing two social sensing mechanisms.

2.3.1 Centralized Social Sensing

Centralized social sensing means that the users send their information to a single destination via the OSN application. This destination is the server of the OSN application. This can be regarded as many-to-one communication. In this social sensing scheme, the source signal is directed to the central server of the OSN application. Users in a sensing proximity of an event send their observations to the common receiver, which also resembles the common sink or fusion center in WSNs.

There are various examples of centralized social sensing schemes in the literature triggered by the heavy use of cell phones for socializing, in addition to communication purposes. For example, mobile phones are utilized for the sensing applications in [13, 14]. The sensors on the phones are the key components for the mobile phone sensing. Mobile phones are equipped with enhanced sensors and strong computational capabilities. A seminal paper, which proposes "Mobile Millennium" [15], uses mobile phone sensing with a collaborative effort with the users, and the extraction of data is conducted according to this effort. It mainly utilizes the GPS information of the mobile phones in the vehicles and processes the traffic information accordingly. It also informs the drivers about the traffic congestion in real time through this traffic monitoring system. Furthermore, participatory sensing applications are also developed to extract useful information from the users of the applications, in which mobile phones form a sensor network providing the ability to gather, share, and analyze [16]. *Micro-Blog* [17] is used to share multimedia content with GPS information to monitor the world and to make queries about the places that have enough information. It is also envisioned that Micro-Blog can be used to enhance carpooling possibilities. According to these approaches, mobile sensing and crowd sensing are the enabling technologies for social sensing in vehicular networks. These approaches lead to the collection of information from the users at the server of the applications, as in centralized social sensing.

In vehicular networks, the centralized social sensing is realized if the users get connected to the server of the social sensing application. Centralized social sensing is the collection of the overall information from the users in vehicles and the estimation of the observed phenomenon using MBD. The information flow from the vehicles to the base station is possible for the connection to the server if there is communication between the vehicle and the base station. The communication for the centralized social sensing is realized with the vehicle-to-infrastructure (V2I) communication, which is explained as follows.

2.3.1.1 Vehicle-to-Infrastructure Communications

The social sensing of an event requires the immediate sharing of the data produced by the users or the vehicles. Hence, access to the Internet is important to convey the information to the server to estimate the phenomenon. Social sensors in the vehicles must access the Internet via Internet gateways through base stations.

The most important enabler for V2I communication is IEEE 802.16j. In the network architecture, there are relay stations, which are actually buses providing access to the vehicles on the roads, and roadside base stations providing Internet access to the relay stations and vehicles in their coverage area [18, 19]. Hence, the users in the vehicles can access the Internet and use social networking services.

IEEE 802.16j enables multihop communication, which also increases the coverage of the networks [20], and this makes IEEE 802.16j a perfect candidate for the communication in centralized social sensing. It enables the collection of data from the users in the vehicles or the sensory information from the vehicles. It also improves the coverage and capacity.

2.3.1.2 Applications for Centralized Social Sensing

As OSNs become ubiquitous, numerous applications have emerged. Twitter is one of the most popular applications of OSNs. The users may communicate with each other via sending tweets. Apart from peer communications, the collective observations of the users about an event are stored in the database of Twitter. These observations can be utilized to estimate the social events. For instance, an accident can be detected by the tweets of the users on the road on which the accident occurs, and the vehicles in the area may select alternative roads to avoid congestion. This is an example in the vehicular domain. The events in the Arab Spring can be regarded as another example. These are examples of centralized social sensing in OSNs. Furthermore, the network, which is formed by the Twitter users in vehicles, can be regarded as an example of a VSSN.

Thanks to next-generation cars, the roads will be mostly equipped with the Internet to enable the social aspects of communication [21]. Hence, the social sensors in vehicles have the ability to send their observations and readings with information dissemination techniques in vehicular networks. The estimation of an event signal is the main goal of social sensing. Hence, the transmission of packets of the users to the server for the estimation of an observed signal is of utmost importance.

The following applications may be considered to be helpful in understanding the centralized social sensing in VSSNs. A sensor networking paradigm is employed for not only monitoring but also sensing social events via user-centric applications. To this extent, Miluzzo et al. present a new application called CenceMe [8]. This application is based on mobile phones, and the activity of people is inferred by the sensors in the phones. The social presence is shared by online social services such as Facebook. This application is for general purposes; however, it can be applied to the vehicular network domain to enable sensors in vehicles. Hence, this application can be utilized for social sensing and extending the sensor capabilities of mobile phones with the sensor capabilities of vehicles. Mobile phones are being widely used in social networking. WatchMe [22] uses mobile phones to estimate the motion of its users and the talking activity of the users to determine the form of communication accurately. The form of communication may be text or audio. If this application is used in vehicles, apart from the social perspective, it helps to decrease the accidents of drivers by determining accurate communication forms during driving.

Real-time event extraction for driving information is an application of centralized social sensing [4]. This method uses social sensors' observations to predict events. To this end, the users in the vehicles use social media to report traffic jams, weather conditions on the road, and sensory information from the cars. The authors in [4] state that the driving information is extracted from the text of social media messages of the social sensors. The geographical information is very important for the driving information, and the authors propose a method to estimate the location of the driving information by extracting the geographic coordinates from the terms related to the geographic location in the messages. This method relies on the fact that the people on the road make posts on social media that are related to the events nearby.

Another application of centralized social sensing is to predict the location and intensity of earthquakes [5]. The authors in [5] consider that Twitter users are the sensors, which provide sensory information to other users and the Twitter database. They named the Twitter users as the social sensors. Although these examples are for general applications, the idea of using Twitter for sensing a social event can be used for the VSSN domain.

The information may be aggregated from different social platforms and cross-space heterogeneous communities [6]. To this end, the authors in [6] have proposed a new research area, named cross-community sensing and mining. It brings new research challenges, such as data collection and integration from different communities, and knowledge transfer between different platforms. This will be the case for the VSSN domain. Vehicle social sensors may use different platforms for their collected data. Hence, [6] can be used to handle cross-space data. This cross-space data can be utilized to extract the event features.

Social sensing is also implemented using the sensors of smartphones in [8]. The authors in [8] use social platforms to extract data from the activities of the mobile device users. Furthermore, the authors in [7] extend the vision of social sensing by considering the broader information from the users, which includes posts from Twitter, updates in Facebook, and so forth. Social sensing is used for different contexts in different applications, such as by Amazon to offer new products to its users by considering their previous purchases and product searches [23] and by Google to extract trends from the Google Search results [24]. Furthermore, social sensors are used for pervasive computing; social networks and pervasive networks are merged together as in [7].

Table 2.1 outlines the applications for centralized social sensing, their approaches, and how they can be used for VSSNs.

2.3.2 Distributed Social Sensing

Decentralized social networks are proposed to overcome the negative aspects and problems of the client–server model of centralized social networks [25]. These problems are Big Data mining, privacy and data piracy, and so forth. Decentralized online social networks (DOSNs) offer peer-to-peer architecture. They support direct exchange of messages instead of traversing the server, as in the case of traditional centralized OSNs. They also promote the local connectivity for social networking without Internet access. From this point of view, distributed social sensing can be build up on top of DOSNs decentralized online social networks to extract information about an event in an ad hoc manner.

Table 2.1   Applications for Centralized Social Sensing

Applications	Used Approaches	Potential Use for VSSNs
[8]	Mobile phone sensing	Conveying the observed data to the server
[22]	Crowd sensing	Decreasing the rate of traffic accidents
[4]	Real-time event extraction	Making posts related to events in traffic
[5]	Mobile phone sensing	Providing sensory information to OSN servers
[6]	Cross-community sensing and mining	Integrating different platforms
[7]	Sensors in smartphones	Merging the different contexts from different applications

Decentralized social networks are designed mostly for integrity. *DECENT*, in [26], proposes decentralized social networks to provide security and privacy for the users. Untrusted users can be determined by keeping distributed hash tables. The authors in [27] state that the decentralization of social networking will provide more freedom to exchange information. The users have more control of their data by this approach. These approaches can be used to realize distributed social sensing to extract information from the events sensed by users on the roads.

Distributed social sensing means that the users in a specific group exchange social information about an event, and the event is estimated by the information provided by neighboring users only. In this type of sensing, the users extract information by themselves from the collected data. To this end, the users on the roads can communicate with each other to socialize and share information about the conditions of the road. For instance, a user in a vehicle may inform other users about the congestion or accident on the road it previously used. The possible scheme for this type of communication is vehicle-to-vehicle communication.

2.3.2.1 Vehicle-to-Vehicle Communications

Intervehicle communication provides communication between the drivers on the roads either via line of sight or via multiple hops through other vehicles between them [28]. The vehicles on the roads form mobile ad hoc networks (MANETs). Vehicular ad hoc networks (VANETs) are subset of MANETs. They communicate with each other in a distributed manner apart from the centralized communication via the cellular backbone. The intervehicle communication is mainly used for safe driving, as in [29]. To this end, the authors in [29] state that the applications of intervehicular communication may be road information dissemination between drivers and coordination between vehicles to avoid collisions. On the other hand, V2V communication may also be used for social sensing due to the proliferation of social networks. Distributed social sensing aims to integrate V2V communication with online social networking services.

V2V communication is realized through the onboard unit (OBU), which is mounted on vehicles to enable communication between them [30]. It contains a network device based on a IEEE 802.11p radio technology. The connection is established through a IEEE 802.11p communication channel.

V2V communication supports ad hoc communication, which offers instant messaging and information exchange for social updates. Integrated sensors, GPS receivers, and network interface help in collecting, processing, and disseminating data [30]. Hence, V2V communication actually does not need social media applications; it only needs a communication link. As

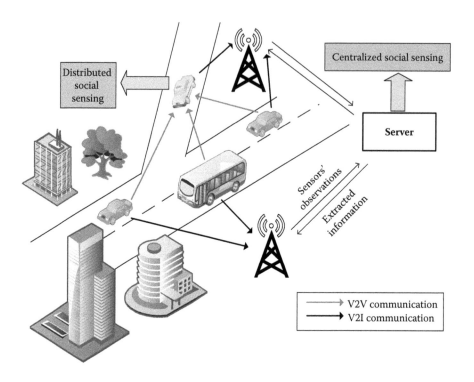

Figure 2.3   V2V and V2I communications.

seen in Figure 2.3, V2V communications provide direct communication between vehicles. On the other hand, V2I communication necessitates base stations to send the information to the server.

2.3.2.2   Applications for Distributed Social Sensing

V2V communication brings a new perspective to the social networks. Social driving, which makes it possible to have social bindings between neighboring drivers with common interests, is realized by V2V communications. According to Type 1 (general information service) and Type 2 (safety information service) applications stated in [31], information about the weather, road and traffic conditions, and road safety can be queried from other drivers. Type 1 and Type 2 vehicular applications based on information services can perform information exchange. Furthermore, intervehicle communication offers a great number of advantages apart from social networking, including improvement of the vision of drivers and road safety [31]. Furthermore, V2V communication, that is, intervehicular communication, is an important component of the intelligent transport system (ITS), which provides management of traffic.

Distributed social sensing is possible by extraction of information from the data provided by the vehicles in a neighborhood. The drivers can share their findings on the road to inform each other about the situations they encounter. To this end, the users in a vehicle can extract information about the road conditions, traffic volume, file sharing, and entertainment. The users on the road are social neighbors who can interact via V2V communication techniques. Social interactions between the drivers contribute to distributed social sensing.

A new social sensing mechanism, named *SmallBlue* [32], is proposed to increase the efficiency between employees in a company, decrease duplicate efforts for the same job, and increase time for

Table 2.2 Applications for Distributed Social Sensing

Applications	Potential Advantages for VSSNs
CarTalk200 [29]	Increase in the efficiency Increase of the collaboration between users
Roadspeak [3]	Voice chatting between users in vehicles Forming virtual networks as in VSSNs
MobiSoC [33]	Location-based mobile social connectivity

innovation. It is a social sensing software that enables collaboration between groups of people in an office. This social sensing mechanism can be used in vehicular social sensing such that a system can be implemented enabling the vehicles to share information while avoiding duplicate information.

Voice chatting is an important application of social networking in vehicular communication. This application was first proposed by the authors in [3] and is named *RoadSpeak*. Roadway commutes usually use the same roads in the same time frame. This situation offers a virtual vehicular network, which can also be regarded as a VSN. It allows the drivers to join and communicate with the other drivers on the road. According to [3], the virtual networks are formed for entertainment, utility, and emergency purposes. Although RoadSpeak utilizes 3G communications system, V2V communication can be utilized to make the vehicular network decentralized. From the perspective of social sensing, each driver can extract information from the conservation with other drivers on the road, which enables social sensing in VSNs. Furthermore, a middleware, named *MobiSoC* [33], is proposed to enable location-based mobile social connectivity. This can be used in vehicular networks, which can be regarded as a basis for social sensing on roads.

The above approaches are not designed for vehicular social sensing; however, the proposed methods can be utilized to realize VSSNs. Table 2.2 shows the proposed methods and their advantages to realize vehicular social sensing.

2.4 Sensor-Related Factors in Reliability of Vehicular Social Sensing

OSN users provide observation signals about the source event. They are *social sensors*, which observe and sample the event through their perceptions based on the sensory inputs from the outside world, and perform different behavior patterns depending on their state of mind. The drivers or the passengers in the vehicles are the social sensors for VSSNs. Depending on the application type, they communicate via communication infrastructure or not. Furthermore, the behavior of the user is different. The mobility of the users is also confined to the roads; they are not free to move in the network in a totally independent manner.

Reliability of the event estimation in VSSNs, that is, vehicular social sensing, depends on a number of sensor-related factors. They are explained in a detailed manner in the following subsections.

2.4.1 Sensor Perceptions

Users observe the source signal through their sensory mechanisms that can sense auditory, visual, tactile, olfactory, and gustatory stimuli. The perception, defined as *the process of recognizing and interpreting sensory stimuli*, creates an interpretation of the sensed signal, and the users share these interpreted signals in the OSN. Users' perceptions may alter the observation signal quality. This change is a source of sensor noise in VSSNs. For instance, users with visual impairments may provide noisier observation signals in the estimation of an event observed through visual stimuli.

The vehicles are equipped with sensors; hence, the vehicle users may use these sensors to sense the event. These events may be the road conditions, since modern cars have sensors that report the road conditions, such as icy, slushy, or wet. Therefore, the perception of the mobile users may be through the sensor on the vehicles. The sensing quality of the sensors on vehicles directly affects the sensory data shared with the other users in VSSNs. The drivers perceive the road conditions by these sensors; hence, a malfunction of these sensors may deteriorate the perception of the drivers.

2.4.2 Sensor Behaviors

The reliability of the social sensors heavily depends on users' behaviors. Social learning theory is useful in defining user behaviors [34]. For example, most users aim to learn the source event in social sensing. Therefore, we may assume that participating users are confronted with sharing their true observations. However, deviant behavior, that is, the actions that adversely affect the overall sensing, is frequently observed in OSNs. Some users may alter the performance of the social sensing by providing false information in different levels, depending on their presence in the community.

2.4.3 Sensor Samples

OSNs provide different post types for information sharing; for example, while Twitter mostly allows 140-character limited text messages, Instagram allows images and short-duration videos. Moreover, Facebook allows posts that include texts, images, web links, or videos. The data type determines the representation of user observations, which is encoding of the user observation signal as shown in Figure 2.2. Information can be represented in many forms, and the forms may introduce different levels of error while encoding [35]. Furthermore, in distributed social sensing, the sensor samples can be text, video, and audio via peer-to-peer links.

Social sensors sample the source signal through their perceptions and behaviors, and encode their signals to share them in the OSN. Therefore, the main open issue regarding social sensors is their characterization. They can be itemized as follows:

- OSN users are social sensors with different noise characteristics limited by their perceptive capabilities. Therefore, noise processes included in samples of social sensor nodes should be modeled in social sensing using social learning theories on human perception.

- Social sensors may perform different behavior patterns while sharing their observation signals. These behavior patterns, for example, deviant behavior, should be investigated and modeled to observe the effect of user behaviors on user observation signals.

- Information-bearing capabilities of the data types may vary for different source signal characteristics. Therefore, understanding the impact of different representations of information is essential in social sensing. The different representations of observation signals should be modeled, and error processes in encoding should be studied.

2.5 Characterization of Mobile Big Data in Vehicular Social Sensing

The ultimate goal of VSSNs is to estimate a phenomenon based on sensory data collected in the network, that is, MBD obtained from vehicular social sensor nodes. However, MBD poses challenges for vehicular social sensing due to its volume, variety, and velocity characteristics. Given the social sensing or any Big Data application, it may be required that the algorithm should be accurate within a distortion bound, scalable for varying amounts of Big Data, and efficient in terms of data storage level and computing time [36].

In addition to the characteristics of social sensors described in Section 2.4, it may be advantageous to further investigate the network characteristics of VSSNs for social sensing, as mobile social sensor nodes and their observations are dependent on social, temporal, and spatial relationships. First, the users with in VSSNs and also the users between different VSSNs may have friendship relationships that can make their observations dependent. Second, due to the nature of the source signal, that is, physical phenomenon, user observations are likely to be dependent on temporal and spatial domain. Using a sensor networking approach, these characteristics of MBD in VSSNs may be leveraged to achieve maximum efficiency in social sensing and also other data mining applications.

2.5.1 *Spatiotemporal Correlation*

Each mobile social sensor node in VSSNs shares its observation at a specific time and location, and temporal information is the key element of MBD, as the user's post is stamped with time information when it is created by the user. On the other hand, spatial information, that is, the location of the OSN user while sharing data, is additional information whose availability is determined by the user. For example, if the user chooses to mention its location via geotag, a location tagging feature that appears in most OSNs, then its location information is publicly available, in addition to the post's content. Some studies show that users most of the time do not use geotags in their observations [37], yet there are several methods to extract location information by using their contents [38].

In most social sensing scenarios, for example, disaster response, traffic speed estimation, or weather forecasting, users are densely located in the vicinity of the event region. Therefore, their observations might be highly correlated in the time or space domain. Thus, investigation of spatiotemporal relationships in VSSNs may be beneficial in event prediction with VSSNs. In several studies, spatiotemporal information is utilized for event estimation. For example, [39] develops a deep learning based traffic flow prediction method, which accounts for spatial and temporal correlations. The authors in [40] study earthquake prediction with Twitter. In the study, sentiment analysis is performed on the tweets, and a probabilistic spatiotemporal model, which can estimate the center location of the source signal, that is, the earthquake, is developed.

On the other hand, spatiotemporal characteristics of user observations can be exploited to handle challenges in MBD through effective sampling. For example, [41] introduces correlation-based medium access control and correlation-based reliable event transport techniques in WSNs. The techniques may determine the selection of sensor nodes based on their spatial relationships; for example, sensors with more spatial separation are likely to contribute less correlated data to the sink and decrease the level of bias. In addition, temporal characteristics of the source signal can be exploited to minimize the energy expenditure in the WSN by utilizing the reporting frequency of the nodes, that is, time separation between the samples of each sensor node [41]. In VSSNs, sensor nodes mostly share information in data forms in a text, image, or video depending on the OSN application; however, spatiotemporal characteristics of the source signal can still be leveraged for efficient data sampling and source selection in MBD.

2.5.2 *Social Intercorrelation*

In addition to spatiotemporal characteristics of the source signal, sensor nodes are generally connected to each other through friendship relationships in VSSNs. A friendship relationship may introduce bias in social sensing since social nodes that are friends with each other in the VSSNs may share correlated observations. For instance, the friends in VSSNs may be the commuters using the same road in the same time frame.

Social interactions are widely studied in social sciences, including development economics, sociology, and marketing, through social intercorrelation, which is a parameter describing the level of dependence between the users in the social network. In social sensing, social intercorrelation is also investigated by using social network analysis, that is, by determining the connectivity of the users in the social network. For example, [42] suggests that independent sources are likely to provide more accurate data and investigates algorithms for diversifying source selection in Twitter for social sensing applications. It develops techniques that maximize the total independence of the source nodes in the network based on the social distance defined by the social graph. For instance, if two nodes in a directed social graph do not have any way to interact with each other, then these nodes are assumed to be independent. The results of the study show that diversified sources are more efficient in terms of the number of nodes, that is, source selection, and the accuracy of estimation [42]. The idea of source diversification can be extended to MBD, however, MBD; introduces additional challenges due to the variety of VSSNs and the interactions between the OSN users.

Broadly, we identify key open issues in the exploitation of spatiotemporal and intersocial correlations in MBD as follows:

■ For varying source signal characteristics in MBD applications, spatiotemporal relationships between user observation signals in the VSSNs should be analyzed.
■ Social intercorrelation should be modeled within and across the VSSNs using social network analysis methods and social interaction models depending on the VSSN features.
■ Effective source selection and data sampling methods, which exploit the spatiotemporal and social characteristics of the user observations and the source signal, should be developed for MBD in VSSNs.

2.6 Conclusion

In this chapter, we introduce a novel network paradigm, named vehicular social sensor networking, for the first time in the literature. We point out that the integration of OSNs with vehicular sensor networks has resulted in this networking paradigm. In this model, mobile users become key contributors of the data production due to events in VSNs, in addition to sensors embedded in vehicles. This paradigm aims to extract information from the observation of the users in vehicles. We investigate fundamentals of the sensor networking approach in social networks to realize social sensing in VSNs, and to determine the different networking approaches and the key elements in a VSSN. We introduce two social sensing mechanisms: centralized and distributed social sensing. In centralized social sensing, all the observation packets are collected in a server and form MBD. On the other hand, in distributed social sensing, social sensors form peer-to-peer networks and the event extraction is possible by the observation packets collected from the neighbor social sensors. Furthermore, we investigate the most important factors influencing the reliability of the event estimation for VSSNs, one of which is the mobile user behavior. The effects of spatiotemporal correlations and social intercorrelations are examined for the sensor networking approach to utilize MBD generated in VSSNs.

References

1. Danah M. Boyd and Nicole B. Ellison. Social network sites: Definition, history, and scholarship. *Journal of Computer-Mediated Communication*, 13(1):210–230, 2007.

2. Julia Heidemann, Mathias Klier, and Florian Probst. Online social networks: A survey of a global phenomenon. *Computer Networks*, 56(18):3866–3878, 2012.

3. Stephen Smaldone, Lu Han, Pravin Shankar, and Liviu Iftode. Roadspeak: Enabling voice chat on roadways using vehicular social networks. In *Proceedings of the 1st Workshop on Social Network Systems*, pp. 43–48. New York: ACM, 2008.

4. Takeshi Sakaki, Yutaka Matsuo, Tadashi Yanagihara, Naiwala P. Chandrasiri, and Kazunari Nawa. Real-time event extraction for driving information from social sensors. In *2012 IEEE International Conference on Cyber Technology in Automation, Control, and Intelligent Systems (CYBER)*, pp. 221–226. Piscataway, NJ: IEEE, 2012.

5. Takeshi Sakaki, Makoto Okazaki, and Yutaka Matsuo. Earthquake shakes twitter users: Real-time event detection by social sensors. In *Proceedings of the 19th International Conference on World Wide Web*, pp. 851–860. New York: ACM, 2010.

6. Bin Guo, Zhiwen Yu, Daqing Zhang, and Xingshe Zhou. Cross-community sensing and mining. *IEEE Communications Magazine*, 52(8):144–152, 2014.

7. Alberto Rosi, Marco Mamei, Franco Zambonelli, Simon Dobson, Graeme Stevenson, and Juan Ye. Social sensors and pervasive services: Approaches and perspectives. In *2011 IEEE International Conference on Pervasive Computing and Communications Workshops (PERCOM Workshops)*, pp. 525–530. Piscataway, NJ: IEEE, 2011.

8. Emiliano Miluzzo, Nicholas D. Lane, Kristóf Fodor, Ronald Peterson, Hong Lu, Mirco Musolesi, Shane B. Eisenman, Xiao Zheng, and Andrew T. Campbell. Sensing meets mobile social networks: The design, implementation and evaluation of the cenceme application. In *Proceedings of the 6th ACM Conference on Embedded Network Sensor Systems*, pp. 337–350. New York: ACM, 2008.

9. Kardelen Cepni and Ozgur B. Akan. Social sensing model and analysis for event detection and estimation with Twitter. In *2014 IEEE 19th International Workshop on Computer Aided Modeling and Design of Communication Links and Networks (CAMAD)*, pp. 31–35. Piscataway, NJ: IEEE, 2014.

10. Rita Tan Sim Tse, Dawei Liu, Fen Hou, and Giovanni Pau. Bridging vehicle sensor networks with social networks: Applications and challenges. In *IET International Conference on Communication Technology and Application (ICCTA 2011)*, pp. 1–5. Stevenage, UK: IET, 2011.

11. Anna Maria Vegni and Valeria Loscri. A survey on vehicular social networks. *IEEE Communications Surveys & Tutorials*, 17(4):2397–2419, 2015.

12. Tad Hogg and Lada Adamic. Enhancing reputation mechanisms via online social networks. In *Proceedings of the 5th ACM Conference on Electronic Commerce*, pp. 236–237. New York: ACM, 2004.

13. Nicholas D. Lane, Emiliano Miluzzo, Hong Lu, Daniel Peebles, Tanzeem Choudhury, and Andrew T. Campbell. A survey of mobile phone sensing. *IEEE Communications Magazine*, 48(9):140–150, 2010.

14. Raghu K. Ganti, Fan Ye, and Hui Lei. Mobile crowdsensing: Current state and future challenges. *IEEE Communications Magazine*, 49(11):32–39, 2011.

15. Nokia Research Center UC Berkeley and NAVTEQ. Mobile millennium. http://traffic.berkeley.edu.

16. Jeffrey A. Burke, Deborah Estrin, Mark Hansen, Andrew Parker, Nithya Ramanathan, Sasank Reddy, and Mani B. Srivastava. Participatory sensing. Los Angeles: Center for Embedded Network Sensing, 2006.

17. Shravan Gaonkar, Jack Li, Romit Roy Choudhury, Landon Cox, and Al Schmidt. Micro-blog: Sharing and querying content through mobile phones and social participation. In *Proceedings of the 6th International Conference on Mobile Systems, Applications, and Services*, pp. 174–186. New York: ACM, 2008.

18. Valeria Loscri. A queue based dynamic approach for the coordinated distributed scheduler of the IEEE 802.16. In *IEEE Symposium on Computers and Communications 2008 (ISCC 2008)*, pp. 423–428. Piscataway, NJ: IEEE, 2008.

19. V. Loscri and G. Aloi. Transmission hold-off time mitigation for IEEE 802.16 mesh networks: A dynamic approach. In *Wireless Telecommunications Symposium 2008 (WTS 2008)*, pp. 31–37. Piscataway, NJ: IEEE, 2008.

20. Vasken Genc, Sean Murphy, Yang Yu, and John Murphy. IEEE 802.16j relay-based wireless access networks: An overview. *IEEE Wireless Communications*, 15(5):56–63, 2008.

21. Georgios Karagiannis, Onur Altintas, Eylem Ekici, Geert Heijenk, Boangoat Jarupan, Kenneth Lin, and Timothy Weil. Vehicular networking: A survey and tutorial on requirements, architectures, challenges, standards and solutions. *IEEE Communications Surveys & Tutorials*, 13(4):584–616, 2011.

22. Natalia Marmasse, Chris Schmandt, and David Spectre. WatchMe: Communication and awareness between members of a closely-knit group. In *UbiComp 2004: Ubiquitous Computing*, pp. 214–231. Berlin: Springer, 2004.

23. Cai-Nicolas Ziegler, Georg Lausen, and Joseph A Konstan. On exploiting classification taxonomies in recommender systems. *AI Communications*, 21(2–3):97–125, 2008.

24. Hal R. Varian and Hyunyoung Choi. Predicting the present with Google trends. Google Research Blog, 2009. http://googleresearch.blogspot.com/2009/04/predicting-present-with-google-trends.html

25. Anwitaman Datta, Sonja Buchegger, Le-Hung Vu, Thorsten Strufe, and Krzysztof Rzadca. Decentralized online social networks. In *Handbook of Social Network Technologies and Applications*, pp. 349–378. Berlin: Springer, 2010.

26. Sonia Jahid, Shirin Nilizadeh, Payal Mittal, Nikita Borisov, and Apu Kapadia. Decent: A decentralized architecture for enforcing privacy in online social networks. In *2012 IEEE International Conference on Pervasive Computing and Communications Workshops (PERCOM Workshops)*, pp. 326–332. Piscataway, NJ: IEEE, 2012.

27. Ching-man Au Yeung, Ilaria Liccardi, Kanghao Lu, Oshani Seneviratne, and Tim Berners-Lee. Decentralization: The future of online social networking. In *W3C Workshop on the Future of Social Networking Position Papers*, Vol. 2, pp. 2–7, Barcelona, 2009.

28. Jun Luo and Jean-Pierre Hubaux. A survey of inter-vehicle communication. Technical report. EPFL, Lausanne, Switzerland, 2004.

29. Dirk Reichardt, Maurizio Miglietta, Lino Moretti, Peter Morsink, and Wolfgang Schulz. Cartalk 2000: Safe and comfortable driving based upon inter-vehicle-communication. In *IEEE Intelligent Vehicle Symposium 2002*, Vol. 2, pp. 545–550. Piscataway, NJ: IEEE, 2002.

30. Saif Al-Sultan, Moath M Al-Doori, Ali H Al-Bayatti, and Hussien Zedan. A comprehensive survey on vehicular ad hoc network. *Journal of Network and Computer Applications*, 37:380–392, 2014.

31. Theodore L. Willke, Patcharinee Tientrakool, and Nicholas F. Maxemchuk. A survey of inter-vehicle communication protocols and their applications. *IEEE Communications Surveys & Tutorials*, 11(2):3–20, 2009.

32. Ching-Yung Lin, Kate Ehrlich, Vicky Griffiths-Fisher, and Christopher Desforges. Smallblue: People mining for expertise search. *IEEE MultiMedia*, 15(1):78–84, 2008.

33. Ankur Gupta, Achir Kalra, Daniel Boston, and Cristian Borcea. MobiSoC: A middleware for mobile social computing applications. *Mobile Networks and Applications*, 14(1):35–52, 2009.

34. Ronald L. Akers and Wesley G. Jennings. Social learning theory. In Mitchell Miller, ed., *The Oxford Handbook of Innovation*, pp. 266–290. Thousand Oaks, CA: SAGE, 2009.

35. Paul A. Kolers and William E. Smythe. Symbol manipulation: Alternatives to the computational view of mind. *Journal of Verbal Learning and Verbal Behavior*, 23(3):289–314, 1984.

36. Xindong Wu, Xingquan Zhu, Gong-Qing Wu, and Wei Ding. Data mining with big data. *IEEE Transactions on Knowledge and Data Engineering*, 26(1):97–107, 2014.

37. Bumsuk Lee and Byung-Yeon Hwang. A study of the correlation between the spatial attributes on Twitter. In *2012 IEEE 28th International Conference on Data Engineering Workshops (ICDEW)*, pp. 337–340. Piscataway, NJ: IEEE, 2012.

38. Kisung Lee, Raman Ganti, Mudhakar Srivatsa, and Prasant Mohapatra. Spatio-temporal provenance: Identifying location information from unstructured text. In *2013 IEEE International Conference on Pervasive Computing and Communications Workshops (PERCOM Workshops)*, pp. 499–504. Piscataway, NJ: IEEE, 2013.

39. Yisheng Lv, Yanjie Duan, Wenwen Kang, Zhengxi Li, and Fei-Yue Wang. Traffic flow prediction with big data: A deep learning approach. *IEEE Transactions on Intelligent Transportation Systems*, 16(2):865–873, 2015.

40. Rui Li, Kin Hou Lei, Ravi Khadiwala, and Kevin Chen-Chuan Chang. Tedas: A Twitter-based event detection and analysis system. In *2012 IEEE 28th International Conference on Data Engineering (ICDE)*, pp. 1273–1276. Piscataway, NJ: IEEE, 2012.

41. Mehmet C Vuran, Özgür B Akan, and Ian F Akyildiz. Spatio-temporal correlation: Theory and applications for wireless sensor networks. *Computer Networks*, 45(3):245–259, 2004.

42. Md. Yusuf S. Uddin, Md. Tanvir Al Amin, Hieu Le, Tarek Abdelzaher, Boleslaw Szymanski, and Tommy Nguyen. On diversifying source selection in social sensing. In *2012 Ninth International Conference on Networked Sensing Systems (INSS)*, pp. 1–8. Piscataway, NJ: IEEE, 2012.

DATA DISSEMINATION IN VSNs

Chapter 3

Social Evolving Graph-Based Connectivity Model for Vehicular Social Networks

Mahmoud Hashem Eiza
School of Physical Sciences and Computing, University of Central Lancashire, Preston, United Kingdom

Qi Shi
Department of Computer Science Liverpool, John Moores University, Liverpool, United Kingdom

Contents

3.1 Introduction

Nowadays, social networking over the Internet has become one of the most popular methods for social interactions among people thanks to the modern and ubiquitous communication technologies and devices. Besides the traditional online social networks, which are offered by service providers such as Facebook, Twitter, and LinkedIn, mobile social networks (MSNs) have emerged as a new platform over which participants interact within a virtual social network using their mobile devices. These mobile devices take advantage of their close proximity and leverage different communication technologies, such as Bluetooth and Wi-Fi Direct. Thus, MSNs offer the possibility of opportunistic social interaction where opportunistic networking is utilized to allow each node to send, receive, and relay information without a server dictating the communications. This feature makes MSNs an attractive option for supporting social interactions and collaborations among people in a number of mobile environments where MSN can take advantage of both infrastructure-based wireless networks, for example, the mobile Internet, and opportunistic networks, such as wireless mobile ad hoc networks [1].

Vehicular social networks (VSNs) are one of the main application domains of MSNs. VSNs are defined as decentralized opportunistic communication networks that facilitate social interactions, including content creating and sharing between travelers on roadways. Due to the lack of high-rate Internet connections on roadways, especially on highways and rural areas, VSNs encourage vehicles' travelers to create, share, and relay information using the available low-cost communication links in vehicular networks, including vehicle-to-vehicle (V2V) and vehicle-to-infrastructure (V2I) communications. Direct inquiry of others with similar experiences in proximity over social networks tends to be the most convenient and efficient approach to acquire up-to-date, specialized, and domain-specific content and information for travelers [2]. Furthermore, a recent TripAdvisor survey of more than 1700 U.S. respondents revealed that 76% of travelers share their travel experience, including photos and clips, via social networks, and 52% do that while traveling or driving back home [3]. Thus, VSNs represent a unique form of localized MSN that exploit the vehicular communication links and offer vehicular travelers the opportunity to engage in social activities along the road.

Given the unique features they provide, VSNs can serve as a platform for various vehicular and traffic-related applications. Therefore, VSNs have received more attention and research efforts from academia and industry worldwide [28]. These efforts have resulted in the development of many applications and frameworks that can operate upon VSNs. RoadSpeak [4], Verse [5], Clique Trip [6], NaviTweet [7], and Toyota Friend [8] are a few examples of these applications.

While VSNs promise a new communication platform for social interactions along the roads, they inherit the connectivity problems that already exist in vehicular networks. These include the high mobility of network nodes and the frequent changes of the network topology. In vehicular networks, the network topology can vary when vehicles change their velocities or lanes. These changes depend on the drivers' behaviors, that is, human factors, and road situations and are normally not scheduled in advance. Here, we assume that vehicles are driven by humans. Self-driving vehicles can be considered a part of a VSN; however, this case needs more investigation and is left for future work. Thus, in other words, the VSN can be defined as a vehicular network that takes the social characteristics of human beings, such as human mobility, human selfish status, and human preferences, into account. Therefore, the current connectivity models, which are designed for vehicular networks, cannot guarantee to capture the social evolving connectivity patterns in VSNs. This problem is the subject of this chapter.

3.2 Basics of Social Theory

As we mentioned in Section 3.1, the human factor has a significant impact on the operations of and, consequently, the performance of VSNs. The human factor in VSNs can be considered from two different points of view: the passengers' social behaviors and the drivers' social behaviors, since driving itself has been constructed as a set of social practices, embodied disposition, cybernetic associations, and physical affordances [9]. However, in this chapter, we consider the social behavior of travelers in general, that is, both drivers and passengers, and the social aspects of vehicles as network nodes in the social network. Considering different behaviors of different travelers in the same vehicle in the context of VSNs is an open research challenge and is left for future work.

In social theory, there are several indexes that can be used to localize the most significant nodes and quantify their relative importance to other nodes [10,11]. These indexes and measures are similar to those utilized in graph theory since the social network itself represents a communication graph. Thus, in this section, we follow the categorization of graph theory to these social indexes and metrics [12], which are concerned with local, network-wide, and community-wide metrics in VSNs. In the following, we describe these measures, along with the model of social morality of vehicular travelers.

3.2.1 Local Metrics

3.2.1.1 Propinquity

Under equal conditions, propinquity means that if two vehicles are geographically near to each other, they are more likely to be connected.

3.2.1.2 Homophyly

In social theory, homophyly is defined as the common social attributes, that is, the similarity, between two users, such as having the same favorites, working for the same organization, and having the same traveling destination. Thus, it is more likely that travelers with the same social attributes, that is, high homophyly, have a connection. Thus, the higher the homophyly, the more likely two vehicles will be socially connected [11].

Let HP_i be the social attributes of an entity n_i, where each item in HP_i is a binary variable that indicates whether n_i has an interest in the corresponding item. For instance, let us assume the following social attributes or interests {Football, Rap music, Thai food, Mountain climbing} and $HP_i = \{1,1,1,0\}$; that is, n_i likes watching football, listening to rap music, and eating Thai food, but he or she does not like mountain climbing. In order to match the similarity between two entities in terms of their social attributes or interests, we adopt the vector space model (VSM) as described in Li et al. [13]. Let $HP_i = \{S_{i,x}\}$, where $x \in 1 \ldots n$ and $S_{\{i,x\}} \in \{0,1\}$ are the social attributes or interests profile of n_i. The homophyly between two entities n_i and n_j, that is, the social interests similarity, can be evaluated as follows [2]:

$$SHP_{ij} = \frac{\sum_{k=1}^{n} S_{i,k} S_{j,k}}{\sqrt{\sum_{q=1}^{n} S_{i,q}^2} \cdot \sqrt{\sum_{y=1}^{n} S_{j,y}^2}} \tag{3.1}$$

If $SHP_{ij} = 1$, then users n_i and n_j have exactly the same social attributes or interests, and consequently, they are more likely to socially communicate, that is, create a social connection in VSNs. Otherwise, if $SHP_{ij} = 0$, then n_i and n_j have no interests in common and it is unlikely for them to have a social connection in VSNs. The evaluation of the homophyly SHP_{ij} factor between two travelers in two different vehicles is the first step to determine the likelihood of establishing a social connection between them in the context of VSN.

3.2.1.3 Degree Centrality

In definition, a central node is the one that relates to a large number of nodes in the network, that is, has a large number of in-links and out-links with other nodes. The degree of a node n_i can be measured by counting the number of links incident to it and is represented by $d(n_i)$ [14]. Since the distinction between in-links and out-links in social networks is not needed, the centrality of a node n_i, denoted as $C_D(n_i)$, can be calculated as follows:

$$C_D(n_i) = d(n_i) = \sum_{\forall i \neq j} x_{ij} \tag{3.2}$$

where $x_{ij} = 1$ if i is incident to j and $x_{ij} = 0$ otherwise. It can be noted that $C_D(n_i)$ depends on the size of the network, and it becomes complex to use when comparing different networks. Let N be the total number of nodes in the network; one way to standardize the degree centrality metric $C_D(n_i)$ is to divide Equation 3.2 by the maximum number of nodes that n_i can be connected to as follows:

$$C'_D(n_i) = d(n_i) = \frac{\sum_{\forall i \neq j} x_{ij}}{(n-1)} \tag{3.3}$$

In the context of VSNs, choosing nodes with a high degree centrality index to forward a message means that the chance of delivering this message to its destination will be high.

3.2.1.4 Social Link Duration

In order to have a social link SL_{ij} between two nodes n_i and n_j, a communication link l_{ij} should first exist; that is, both vehicles should be within the transmission range of each other. Since the social attributes of travelers are less likely to change over the road, that is, their homophyly, the social link duration mainly depends on the communication link duration between two vehicles. Let H denote the wireless transmission range and $v_i(t)$ and $v_j(t)$ the velocities of n_i and n_j at time t, respectively; the social link duration $SL_{ij}(t)$ can be accurately estimated as follows:

$$SL_{ij}(t) = \frac{H - \theta\sqrt{(y_i(t) - y_j(t))^2 + (x_i(t) - x_j(t))^2}}{|v_i(t) - \vartheta v_j(t)|} \tag{3.4}$$

where $\theta = -1$ and $\vartheta = 1$ when n_j overtakes n_i, $\theta = 1$ and $\vartheta = 1$ when n_i moves forward in front of n_j, $\theta = -1$ and $\vartheta = -1$ when n_i and n_j are moving toward each other, and $\theta = 1$ and $\vartheta = -1$ when n_i and n_j are moving away from each other.

However, the calculation of $SL_{ij}(t)$ in Equation 3.4 does not take into consideration the possible changes in vehicles' velocity values. Therefore, we utilize the concept of link reliability, which is introduced in Eiza and Ni [15], to accurately estimate the expected social link duration. The link reliability is defined as the probability that the communication link between two nodes n_i and n_j will stay continuously available over a specified time period. Given $SL_{ij}(t)$, the estimated duration for the continuous availability of a social link SL_{ij} between two vehicles at time t as calculated in Equation 3.4, the link reliability value $r_t(SL_{ij})$ is expressed as follows:

$$r_t(SL_{ij}) = P\{\text{To continue to be available until } t + SL_{ij} | \text{available at } t\}$$

We assume that the velocity of vehicles has a normal distribution [16,17]. Thus, $r_t(SL_{ij})$ can be calculated as follows:

$$r_t(SL_{ij}) = \begin{cases} \int\limits_{t}^{t+SL_{ij}} f(T)dT & \text{if } SL_{ij} > 0 \\ 0 & \text{otherwise} \end{cases} \tag{3.5}$$

where $f(T)$ is defined as follows:

$$f(T) = \frac{4H}{\sigma_{\Delta v}\sqrt{2\pi}} \frac{1}{T^2} e^{-\frac{\left(\frac{2H}{T} - \mu_{\Delta v}\right)^2}{2\sigma_{\Delta v}^2}} \tag{3.6}$$

where $\mu_{\Delta v}$ and $\sigma_{\Delta v}^2$ denote the average value and the variance of relative velocity $\Delta v = |v_i - v_j|$, respectively. Hence, the expected social link duration $ET(SL_{ij})$ can be estimated as follows:

$$ET(SL_{ij}) = r_t(SL_{ij}) \cdot SL_{ij}(t) \tag{3.7}$$

3.2.2 Network-Wide Metrics

3.2.2.1 Closeness Centrality

It can be noticed in Equation 3.2 that the degree centrality metric does not consider the indirect connections that a node can establish with other nodes using the available paths in the network. Thus, the degree centrality metric is not enough to recognize the most important nodes in VSN. We define a node as central-close if it can reach other nodes through short-distance paths. Hence, the closeness centrality metric is related to the inverse of distance between nodes; for example, the higher the distance, the less central-close. In social theory, the shortest-distance path between two nodes is defined as a geodesic. Thus, the closeness centrality index should consider the geodesics that a given node has to all other nodes in the network. Let $d(n_i, n_j)$ be the geodesic between two nodes n_i and n_j; the standardized closeness centrality $C_C'(n_i)$ of a node n_i can be calculated as follows:

$$C_C'(n_i) = \frac{n-1}{\left(\sum\limits_{j=1, i \neq j}^{n} d(n_i, n_j)\right)} \tag{3.8}$$

In Equation 3.8, it can be noticed that the closeness centrality index for n_i will be zero if there is at least one node n_j that is unreachable from n_i; that is, its geodesic will be infinite. In the context of VSNs, choosing nodes with a high closeness centrality index to forward a message will optimize the resources needed to deliver it and ensure a faster delivery as well.

3.2.2.2 Betweenness Centrality

Betweenness is another measure of centrality that focuses on nodes lying in the path between other nodes. In order to calculate the betweenness centrality index, it is assumed that nodes prefer to communicate via the shortest paths in the network. Thus, the standardized betweenness centrality $C_B'(n_i)$ of a node n_i, which expresses the number of shortest paths, that is, geodesics, that pass through n_i, is calculated as follows:

$$C_B'(n_i) = \frac{\displaystyle\sum_{j<k, i\neq j, i\neq k} \frac{g_{jk}(n_i)}{g_{jk}}}{\frac{((n-2)(n-1))}{2}} \tag{3.9}$$

where $g_{jk}(n_i)$ is the number of geodesics linking n_j and n_k that contain n_i in between and g_{jk} is the total number of geodesics linking n_j and n_k. A node with a high betweenness centrality index plays the role of a "broker" and has a great influence on the data dissemination in a VSN.

3.2.2.3 Bridging Centrality

The bridging centrality of a node n_i is expressed as the product of its betweenness centrality and a bridging coefficient $\beta(n_i)$. The bridging centrality metric defines nodes that are central to the network graph, connecting two highly connected subgraphs, and have a low number of direct connections relative to their neighbor connections. The bridging coefficient $\beta(n_i)$ is the ratio of the inverse of a node degree to the sum of the inverses of all its neighbors' degrees. The bridging centrality can be calculated as follows:

$$C_R'(n_i) = C_B'(n_i)\beta(n_i) \tag{3.10}$$

where $\beta(n_i)$ is calculated as follows:

$$\beta(n_i) = \frac{C_D'(n_i)^{-1}}{\left(\displaystyle\sum_{j=1, i\neq j} C_D'(n_j)^{-1}\right)} \tag{3.11}$$

3.2.2.4 Social Path Duration

When two nodes n_i and n_j are not adjacent to each other, the social path SP_{ij} is defined as the set of social links that connect n_i and n_j via multiple hops in the network. Without loss of generality, for any given path SP_{ij}, let us denote the number of its formed links by k, that is, $SP_{ij} = \{SL_1, SL_2, \ldots, SL_k\}$. The expected social path duration $SP_{ij}(t)$ is defined as the minimum of social link durations that comprise this path, that is,

$$SP_{ij}(t) = \min_{SL_\omega \in SP_{ij}} ET(SL_\omega) \qquad \text{where } \omega = 1 \ldots k \tag{3.12}$$

while the reliability of the social path, denoted as $R_t(SP_{ij})$, can be calculated as follows:

$$R_t(SP_{ij}) = \prod_{\omega=1}^{k} r_t(SL_\omega) \tag{3.13}$$

3.2.3 Community-Wide Metrics

3.2.3.1 Number of Clusters

Considering the mobility patterns of vehicles on the roads and the traffic conditions, the formation of nonconnected clusters is inevitable in VSNs. We define the cluster as a subgraph of the whole network that contains a number of connected vehicles where there is a path between any pair of nodes. The higher the number of clusters in VSNs, the lower the chance of creating new social connections among vehicles in the network.

3.2.3.2 Number of Social Groups

Differing from the cluster approach, the social group or community in VSNs is a subgraph of the whole network that still has a connection with the rest of the network. However, the number of intragroup links is larger than the number of intergroup links. Nodes within a social group usually share the same social attributes or interests. The formation of social groups is quite important in VSNs to attract users' attention and encourage them to join. The higher the number of social groups, the better the experience users will have when joining the VSN.

3.2.4 Model of Social Morality

In a fully autonomous system, users independently behave based on the rational calculation of expediency [18]. Generally, users make their decision to act in social interactions from two points of view: (1) economic and motivated by self-interest and (2) noneconomic and motivated by collective interest and moral obligation. In reality, when users violate a deeply internalized norm, which governs users' behavior in economic and noneconomic spheres of activity, they will feel guilty to some extent and would likely punish themselves in some manner, whether anyone else knew of their actions or not. This is known as social morality [19].

   In the context of social networks, both cooperative and noncooperative behaviors of users have a significant impact on the performance of the social network. It has been shown that users who experienced a feeling of guilt after a noncooperative behavior tend to show a higher level of cooperation in the later social interactions [20]. In fact, the feeling of guilt encourages users to depart from their typical noncooperative behavior. In VSNs, the cooperation is highly desirable among users to promote social interactions and, consequently, help with delivering data packets. Since the VSN users are autonomous and intelligent individuals, it is reasonable to assume that they are rational and their behaviors are driven by personal profit and morality. Thus, we are interested in observing two forms of social morality that result based on the decisions made by users to either accept and forward data packets, that is, cooperative behavior, or reject and drop data packets, that is, noncooperative behavior. These two forms of social morality are guilt and high-mindedness, where users feel high-minded when they choose to cooperate and they feel guilty otherwise.

   Let $g(n_i)$ be a self-regulated morality factor for a node n_i, which quantitatively depicts the internal moral force of the user. This factor is based on the following two elements [19]:

1. Morality state $mt(n_i)$. This element has a variable value and reflects the behavior history of the user. It increases by one level for a single cooperation behavior and decreases by one level due to a single defection conduct. The initial state is set to 0, which means neither guilty nor high-minded. States with a positive index are high-minded states that imply frequent cooperative behavior in the past. On the other hand, states with a negative index are guilty states that imply an overwhelming defection conduct in the past.

2. Sociality strength $st(n_i)$. This element is related to the user's personal experience, such as education and habitation. It is a less independent and stabilized measure with short-term behavior changes. If the sociality strength of a user is significant, the user experiences a significant increment of guilt toward a single defection behavior and a significant increment of high-mindedness toward a single cooperation behavior. The $st(n_i)$ value is chosen in the range $[0, 1]$.

In order to evaluate the morality factor $g(n_i)$, the current morality state $mt(n_i)$ and the sociality strength $st(n_i)$ are utilized as follows:

$$g(n_i) = \begin{cases} f(mt(n_i), st(n_i)) & \text{if } mt(n_i) < 0 \\ 0 & \text{otherwise} \end{cases} \qquad (3.14)$$

where the value of $g(n_i)$ increases as $mt(n_i)$ decreases or $st(n_i)$ increases. The function f can be selected from the following three morality functions: the linear function f_1, the natural logarithm function f_e, and the common logarithm function f_{10}.

$$f_1(mt(n_i), st(n_i)) = \delta \cdot st(n_i) \cdot (-mt(n_i))$$

$$f_e(mt(n_i), st(n_i)) = \delta \cdot \ln(1 + st(n_i) \cdot (-mt(n_i))) \qquad (3.15)$$

$$f_{10}(mt(n_i), st(n_i)) = \delta \cdot \log_{10}(1 + st(n_i) \cdot (-mt(n_i)))$$

where δ is a tunable coefficient in the range $[0, +\infty]$. These three morality functions represent three different levels of morality force that affect user cooperation behavior, respectively. They always output a nonnegative value. In the context of VSNs, choosing a node with a high morality factor value is very important to ensure the delivery of data packets. Moreover, this will attract more users to join the VSN and promote the cooperative nature of these networks. This issue is further discussed in Section 3.4.

3.3 Evolving Graph Theory

The evolving graph theory is proposed as a formal abstraction for dynamic networks [21]. The evolving graph is an indexed sequence of λ subgraphs of a given graph, where the subgraph at a given index corresponds to the network connectivity at the time interval indicated by the index number, as shown in Figure 3.1.

It can be observed from Figure 3.1 that edges are labeled with corresponding presence time intervals. For instance, in Figure 3.1, $\{A, D, C\}$ is not a valid journey since edge $\{D, C\}$ exists only in the past with respect to edge $\{A, D\}$. Hence, the journey in the evolving graph is the path in the underlying graph where its edges' time labels are in increasing order. In Figure 3.1, it is easy to find that $\{A, B, E, G\}$ and $\{D, C, E, G\}$ are valid journeys, while $\{D, C, E, G, F\}$ is not.

Let $G(V, E)$ be a given graph and an ordered sequence of its subgraphs, $S_G = G_1(V_1, E_1)$, $G_2(V_2, E_2)$, $G_3(V_3, E_3), \ldots, G_\lambda(V_\lambda, E_\lambda)$, such that $\cup_{i=1}^{\lambda} G_i = G$. The evolving graph is defined as $G' = (S_G, G)$, where the vertices set of G' is $V_{G'} = \cup V_i$ and the edges set of G' is $E_{G'} = \cup E_i$. Suppose that the subgraph $G_i(V_i, E_i)$ at a given index i is the underlying graph of the network during time interval $\mathcal{F} = [t_{i-1}, t_i]$, where $t_0 < t_1 < \cdots < t_\tau$; the time domain $\check{T}$ is now incorporated in the model.

Let Ω be a given path in the evolving graph G' where $\Omega = e_1, e_2, e_3, \ldots, e_k$, with $e_i \in E_{G'}$ in G and $1 \leq i \leq k$. Let $\Omega_\sigma = \sigma_1, \sigma_2, \sigma_3, \ldots, \sigma_k$, with $\sigma_i \in \check{T}$ be the time schedule indicating when

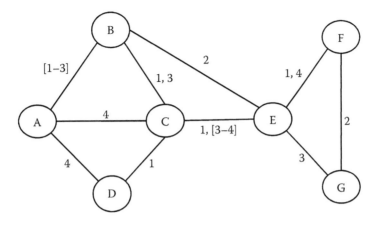

Figure 3.1 Basic evolving graph model. (From Monteiro, J., in *XV Concurso Latinoamericano de Tesis de Maestrìa*, Santa Fe, Argentina, 2008, pp. 1–17.)

each edge of path Ω is to be traversed. We define a journey $J = (\Omega, \Omega_\sigma)$ if and only if Ω_σ is in accordance with Ω, G, and $\mathcal{F}$. This means that J allows the traverse from node n_i to node n_j in G'.

In the current evolving graph theory, three journey metrics are defined: the foremost, shortest, and fastest journeys. They are introduced to find the earliest arrival date, the minimum number of hops, and the minimum delay (time span) path, respectively. Given $J = (\Omega, \Omega_\sigma)$, these are defined as follows:

- The hop count $h(J)$ or the length of J is defined as $h(J) = |\Omega|$.
- The arrival date of the journey $a(J)$ is defined as the scheduled time for the traversal of the last edge in J, plus its traversal time, that is, $a(J) = \sigma_k + t(e_k)$.
- The journey time $t(J)$ is defined as the past time between the departure and the arrival, that is, $t(J) = a(J) - \sigma_1$.

3.4 Social Evolving Graph-Based Connectivity Model for VSNs

3.4.1 *Motivation*

The current evolving graph theory cannot be directly applied to VSNs because the evolving social properties of the VSN communication graph cannot be scheduled in advance. Moreover, the current evolving graph model does not consider the social metrics of the communicating nodes. In order to facilitate the establishment of social links or paths in VSNs and the data forwarding process, we extend the current evolving graph model to develop a social evolving graph–based connectivity (SEGC) model for VSNs. The SEGC model has two main goals in the context of VSNs. First, it captures the social characteristics of the existing nodes, and by considering both social and connectivity metrics, it establishes social links or paths among these nodes. Second, the SEGC model facilitates the data forwarding among the socially connected vehicles using the social theory indexes we have mentioned in Section 3.2, along with the conventional connectivity metrics. In the following, we introduce the proposed SEGC model and explain the data forwarding mechanism that takes advantage of the developed SEGC model.

3.4.2 Social Evolving Graph-Based Connectivity Model

As we have mentioned before, establishing new social connections between two vehicles depends not only on being within the transmission range of each other, but also on their social attributes and interests. Thus, in the proposed SEGC model, each link is characterized with a set of attributes that include all the connectivity and social indexes we have mentioned before. The social link is only established between two vehicles n_i and n_j if it satisfies the following two conditions. First, the SHP_{ij}, that is, the homophyly, should be higher than a predefined threshold Ψ_H; thus, the users share the minimum level of interest. Second, the expected social link duration $ET(SL_{ij})$ should be higher than a predefined time threshold Ψ_L. The value of Ψ_H can be defined or advertised by the vehicle itself; for example, a high value of Ψ_H indicates that the user is only interested in communicating with other users that have a lot in common with himself or herself. On the other hand, the time threshold Ψ_L can be determined by the current application. For instance, in order to share a video clip with other vehicles, the connection time should be long enough to watch or download the video file.

Figure 3.2 shows an example of the SEGC model on a highway at two time instants $t = 0$ s and $t = 5$ s where $\Psi_H = 0.5$ s and $\Psi_L = 10$ s. Each node in Figure 3.2 shows a vehicle on the highway. It can be seen in Figure 3.2 that unlike the corresponding presence time intervals for each link as shown in Figure 3.1, we associate the tuple $(t, SHP_{ij}, ET(SL_{ij}))$ with each link, where t denotes the

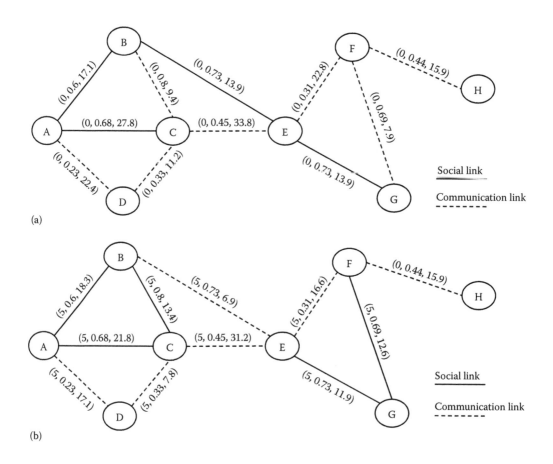

Figure 3.2 SEGC model at (a) $t = 0$ s and (b) $t = 5$ s.

current time, SHP_{ij} denotes the homophyly between n_i and n_j, and $ET(SL_{ij})$ denotes the expected social link duration.

In the SEGC model, the social link between two vehicles is not available if $SHP_{ij}<\Psi_H$ or $ET(SL_{ij})<\Psi_L$. Therefore, even if the communication link exists between two vehicles and satisfies the connection time threshold, for example, the communication link between vehicles A and D in Figure 3.2a, the social link is not established since it does not satisfy the condition of the homophyly as $0.23 < 0.5$. Figure 3.2a shows the SEGC status and the corresponding SHP_{ij} and $ET(SL_{ij})$ values associated with each link at $t = 0$ s. It can be noticed that the following social links are established: $\{A, B\}$, $\{A, C\}$, $\{B, E\}$, and $\{E, G\}$. After 5 s, in Figure 3.2b, the set of the established social links changes and becomes as follows: $\{A, B\}$, $\{A, C\}$, $\{B, C\}$, $\{E, G\}$, and $\{F, G\}$. It is worth noting that all links in Figure 3.2 are eligible to be traversed. However, if the link is eligible to be traversed, it does not necessarily mean that a social link will be established.

In VSNs, we assume that each vehicle along the road has its own version of the SEGC model shown in Figure 3.2. This is possible using the information received within the basic safety messages (BSMs) that are periodically exchanged in vehicular networks when the 5.9 GHz dedicated short-range communication (DSRC) standard is deployed [22]. In this way, each vehicle n_i can be only concerned with vehicles of interest, that is, vehicles that share the same social attributes or interests with n_i. It can be noticed that unlike the conventional evolving graph, the presence time of the social link in the SEGC model is continuous and depends on the current vehicular traffic status and the social attributes of vehicles. In this case, there is no need to check the order of the presence times of the link when searching for a valid journey.

In order to establish a social path, between two nonadjacent vehicles in the network, the same principle is applied. However, forwarding the data along the established multihop social path from the source to the destination should take into account different parameters than those that were considered while establishing the social link. These different parameters are related to the relay vehicles along the established path and are specified in Section 3.4.3.

3.4.3 SEGC-Based Data Forwarding Mechanism

In order to forward data packets in VSNs among nonadjacent connected vehicles, social paths should be established. In this section, we propose a new forwarding data mechanism that can benefit from the SEGC model advantages and properties. The proposed mechanism utilizes the SEGC model and considers both social and connectivity metrics while searching for a path from the source to the destination. The considered metrics are degree centrality, morality factor, closeness centrality, betweenness centrality, and bridging centrality. However, establishing the social path for data forwarding between two vehicles subject to these multiple metrics features a multiconstrained path (MCP) selection, which is proven to be an NP-hard problem [23] if the constraints are mutually independent [24]. Therefore, we propose the following evaluation function $EF(SL_{ij})$ that considers these metrics and its weights:

$$EF(SL_{ij}) = \gamma_g g(n_j) + \gamma_D C'_D(n_j) + \gamma_C C'_C(n_j) + \gamma_B C'_B(n_j) + \gamma_R C'_R(n_j) \qquad (3.16)$$

where γ_g, γ_D, γ_C, γ_B, and γ_R are weighting factors for the morality factor, degree centrality, closeness centrality, betweenness centrality, and bridging centrality, respectively. These factors are chosen in the range [0, 1]. We worked out this function by experimentation, and its validity is illustrated by the simulation results presented later. In order to explain the purpose of these weighing factors, let us assume that the source and destination vehicles belong to different social communities or groups. In this case, γ_B is given a high value because it is important to forward data packets through a vehicle with a high bridging centrality value. On the other hand, γ_g is always given a high value because data packets should be forwarded through vehicles with a high morality factor, that is, vehicles that showed cooperative behavior in the past, to ensure a successful data packet forwarding. When the source vehicle has data to send at time t, it evaluates the communication links in the current SEGC model and assigns each link with a single value $EF(SL_{ij})$, as estimated in Equation 3.16. Finding the optimal path in the SEGC model according to the $EF(SL_{ij})$ value is equivalent to finding the optimal journey in the underlying graph where a modified version of Dijkstra's algorithm can be applied [15]. The modified Dijkstra's algorithm scans all the network nodes in the SEGC model and returns the optimal route according to the $EF(SL_{ij})$ value.

3.5 Performance Evaluation

The main objective of this performance evaluation is to identify the impact of high dynamics of network topology changes in VSNs on the establishment of social connections among the communicating vehicles. In addition, we want to check the benefits of using the proposed SEGC model in the highway scenario. We construct our performance evaluation using the OMNeT++ network simulator [25]. OMNeT++ is an extensible modular component-based C++ simulation library and framework. The simulations are run on a six-lane traffic simulation scenario of a 10 km highway with two independent driving directions in which vehicles move. For each simulation, we perform 20 runs to obtain its average results. The results are compared with those when the SEGC model is not involved; that is, the greedy forwarding mechanism is applied.

In our simulation scenario, the average velocities of the vehicles in the first two lanes are 40 and 60 km/h, respectively, while we change the average velocity of the vehicles in the third lane only, from 60 to 130 km/h. We use the highway mobility model developed in Eiza et al. [26], which is built based on traffic theory rules and considers the drivers' behaviors. The number of vehicles on the highway is 120 and the data packet size is 2 KB. The social attributes profile of each vehicle HP_i is generated randomly to match the following set: {Travel destination, Rap music, Mountain climbing, Thai food, Workplace, Football, Jogging, Cooking}. This set is imaginary and designed for the purpose of this simulation. In a real-world scenario, this set could contain more than 100 elements. The social attributes profile is assumed to be transmitted periodically for vehicles that want to participate in social interactions along the road over VSNs. The morality factor $g(n_i)$ is evaluated using the linear function f_1 in Equation 3.15, where $\delta = 1$. The sociality strength $st(n_i)$ value for each vehicle is randomly selected in the range [0, 1]. The weighting factors in Equation 3.16 are set as follows: $\gamma_g = 1$, $\gamma_D = 0.7$, $\gamma_C = 0.5$, $\gamma_B = 0.5$, and $\gamma_R = 1$ if the source and the destination belong to different social groups; otherwise, $\gamma_R = 0.1$. Finally, the homophyly threshold Ψ_H is randomly selected in the range [0, 1] for each vehicle at the beginning of the simulation run and stays fixed for the rest of the simulation time. The time threshold Ψ_L is set to 10 s. When $SHP_{ij} \geq \Psi_H$ and $ET(SL_{ij}) \geq \Psi_L$ between two vehicles, data packet transmission takes place. The simulation parameters are summarized in Table 3.1.

3.5.1 Performance Metrics

The following performance metrics are considered for the simulations:

■ Packet delivery ratio (PDR): The average ratio of all successfully received data packets at the destination node over all data packets generated by the application layer at the source node.

■ Social connections: The average number of social connections that are established among the communicating vehicles.

Table 3.1 Summary of the Simulation Parameters

Simulation area	1 km × 10 km
Mobility model	Highway
Communication range	450 m
MAC layer	IEEE 802.11p
Vehicle velocities	Normally distributed
Vehicle distances	Exponentially distributed
Number of runs	20
Simulation duration	300 s
Morality function	$f_1(mt(n_i),\ st(n_i)) = \delta \cdot st(n_i) \cdot (-mt(n_i))$
Morality function coefficient	$\delta = 1$
Weighting factors	$\gamma_g = 1,\ \gamma_D = 0.7,\ \gamma_C = 0.5,\ \gamma_B = 0.5,\ \gamma_R = 1$ or $\gamma_R = 0.1$

- Link failures: The average number of communication link failures during the data forwarding process. This metric shows the efficiency of the data forwarding algorithm in avoiding link failures.

- Social path lifetime: The average lifetime of the established social path between two vehicles. A longer lifetime means a more stable and more reliable path.

3.5.2 Simulation Results

In Figure 3.3, it can be seen that the average PDR reduces noticeably when the average velocity in the third lane starts to exceed 80 km/h. This reduction comes from the fact that the network topology becomes more dynamic, and thus links or paths are more vulnerable to disconnection. In this particular case, it is important to establish reliable social paths among the communicating vehicles. Utilization of the SEGC model ensures that only reliable paths are established among the

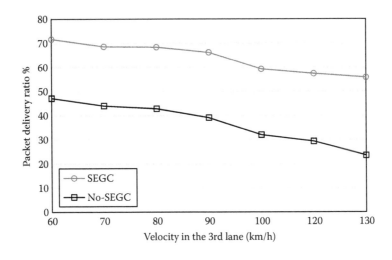

Figure 3.3   Average PDR.

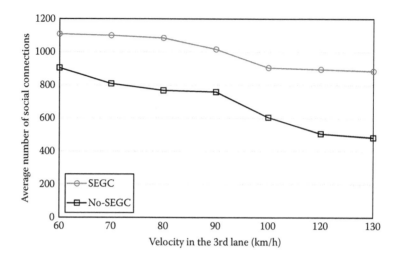

Figure 3.4   Average number of established social connections.

socially connected vehicles. These paths are calculated using the SEGC model where the evolving characteristics of the network topology are considered via Equation 3.13. Moreover, the evaluation function in Equation 3.16 ensures that data packets are relayed through vehicles with a high morality factor and high degree centrality. Thus, the probability of a successful data delivery is high.

The average number of established social connections among the communicating vehicles is shown in Figure 3.4. It can be noticed that when the SEGC model is utilized, the number of social connections is high in comparison with the case where SEGC is not presented. The reason is that each vehicle has its own SEGC model that is updated regularly when a new vehicle enters the communication range of that vehicle. If the homophyly exceeds the defined threshold, the SEGC establishes the social link or path between the two vehicles and commences data packet transmission.

In Figure 3.5, the utilization of SEGC helps in obtaining a very low number of social link failures in comparison with the case when SEGC is not utilized. The number of link failures increases when the velocity increases. In this case, it is essential to accurately capture the changes of vehicular velocities and establish reliable paths between the communicating vehicles. Furthermore, choosing relay vehicles with a high morality factor is crucial to guarantee that the established path will not break if one of the vehicles along it refuses to cooperate in the data packet forwarding process.

In Figure 3.6, we show the average social path lifetime obtained in this performance evaluation. When the SEGC model is utilized, longer social path lifetimes are achieved thanks to establishing the most reliable paths in the network and utilizing the social indexes to forward the data among the communicating vehicles effectively. This observation explains the high PDR shown in Figure 3.3.

3.6 Conclusion

In this chapter, we have extended the evolving graph theory and utilized the social theory concepts to develop a novel SEGC model for VSNs. The proposed connectivity model considers both the social metrics of the communicating vehicles and the conventional connectivity issues in VSNs.

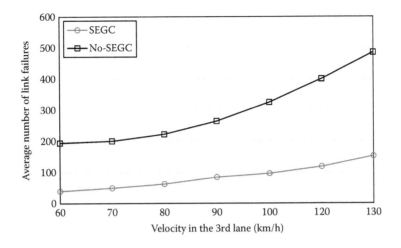

Figure 3.5   Average number of link failures.

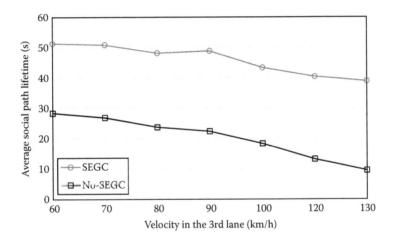

Figure 3.6   Average social path lifetime.

Therefore, the social links or paths are established between the communicating vehicles based on their social characteristics and interests rather than just their kinematic information. The performance of SEGC has been compared with the one where a greedy data forwarding mechanism is utilized through our simulations. The simulation results showed that utilization of the SEGC model helped to achieve a higher PDR and establish stable social paths with longer lifetimes. Since it establishes the most reliable social path between the source and the destination, it also achieves the lowest number of social link failures. The SEGC model shows promising results in the context of VSNs. However, more investigation and therefore more simulations are needed to validate the SEGC model in different traffic scenarios with different traffic parameters. In future work, we will investigate the role of self-driving vehicles and their effects on the connectivity patterns in VSNs. Moreover, we will develop a model to consider different social attribute profiles for individuals inside the vehicles, including the travelers and the drivers.

References

1. Hu, X., T. H. S. Chu, V. C. M. Leung, E. C.-H. Ngai, P. Kruchten, and H. C. B. Chan. 2015. A survey on mobile social networks: Applications, platforms, system architectures, and future research directions. *IEEE Communications Surveys & Tutorials*, pp. 1557–1581.

2. Luan, T. H., R. Lu, X. Shen, and F. Bai. 2015. Social on the road: Enabling secure and efficient social networking on highways. *IEEE Wireless Communications*, vol. 22, no. 1, pp. 44–51.

3. TripAdvisor, Inc. 2012. TripAdvisor survey reveals three quarters of U.S. travelers sharing trip experiences on social networks. September 20. http://www.tripadvisor.co.uk/PressCenter-i5414-c1-Press_Releases.html (accessed June 19, 2015).

4. Smaldone, S., L. Han, P. Shankar, and L. Iftode. 2008. RoadSpeak: Enabling voice chat on roadways using vehicular social networks. In *Proceedings of the 1st Workshop on Social Network Systems*, Glasgow, pp. 43–48.

5. Luan, T. H., X. Shen, F. Bai, and L. Sun. 2015. Feel bored? Join Verse! Engineering vehicular proximity social networks. *IEEE Transactions on Vehicular Technology*, vol. 64, no. 3, pp. 1120–1131.

6. Knobel, M., M. Hassenzahl, M. Lamara, T. Sattler, J. Schumann, K. Eckoldt, and A. Butz. 2012. Clique Trip: Feeling related in different cars. In *Proceedings of the Designing Interactive Systems Conference*, Newcastle, UK, pp. 29–37.

7. Sha, W., D. Kwak, B. Nath, and L. Iftode. 2013. Social vehicle navigation: Integrating shared driving experience into vehicle navigation. In *Proceedings of 14th Workshop on Mobile Computing Systems and Applications*, Jekyll Island, GA, pp. 161–166.

8. Schroeder, S. 2011. Toyota Owners To Get a Private Social Network. May 23. http://mashable.com/2011/05/23/toyota-friend-social-network/ (accessed October 19, 2016).

9. Smyth, T. L., and M. J. King. 2006. Driver–vehicle interactions in 4WDs: A theoretical review. In *Australasian Road Safety Research, Policing Education Conference*, Queensland, Australia.

10. Batallas, D. A., and A. A. Yassine. 2006. Information leaders in product development organizational networks: Social network analysis of the design structure matrix. *IEEE Transactions on Engineering Management*, vol. 53, no. 4, pp. 570–582.

11. Lu, R. 2012. Security and privacy preservation in vehicular social networks. PhD thesis, University of Waterloo, Waterloo, Ontario.

12. Pallis, G., D. Katsaros, M. D. Dikaiakos, N. Loulloudes, and L. Tassiulas. 2009. On the structure and evolution of vehicular networks. In *IEEE International Symposium on Modeling, Analysis & Simulation of Computer and Telecommunication Systems*, London, pp. 1–10.

13. Li, X., L. Guo, and Y. Zhao. 2008. Tag–based social interest discovery. In *Proceedings of the 17th International Conference on World Wide Web*, Beijing, pp. 675–684.

14. Snijders, T., and S. Borgatti. 1999. Non–parametric standard errors and tests. *Connections*, vol. 22, no. 2, pp. 161–170.

15. Eiza, M. H., and Q. Ni. 2013. An evolving graph–based reliable routing scheme for VANETs. *IEEE Transactions on Vehicular Technology*, vol. 62, no. 4, pp. 1493–1504.

16. Niu, Z., W. Yao, Q. Ni, and Y. Song. 2006. Link reliability model for vehicle ad hoc networks. In *London Communication Symposium*, London, pp. 1–4.

17. Schnabel, W., and D. Lohse. 1997. *Grundlagen der Straßenverkehrstechnik und der Verkehrsplanung*. Berlin: Aufl Verlag für Bauwesen.

18. Fukuyama, F. 1996. *Trust: Social Virtues and the Creation of Prosperity*. New York: Free Press.

19. Liang, X., X. Li, T. H. Luan, R. Lu, X. Lin, and X. Shen. 2012. Morality–driven data forwarding with privacy preservation in mobile social networks. *IEEE Transactions on Vehicular Technology*, vol. 61, no. 7, pp. 3209–3222.

20. Ketelaara, T., and W. T. Aub. 2003. The effects of feelings of guilt on the behaviour of uncooperative individuals in repeated social bargaining games: An affect–as–information interpretation of the role of emotion in social interaction. *Cognition and Emotion*, pp. 429–453.

21. Ferreira, A. 2002. On models and algorithms for dynamic communication networks: The case for evolving graphs. In *4e Rencontres Francophones sur les ALGOTEL*, pp. 155–161. Mèze, France.

22. Kenney, J. B. 2011. Dedicated short-range communications (DSRC) standard in the United States. *Proceedings of the IEEE*, vol. 99, no. 7, pp. 1162–1182.

23. Wang, Z., and J. Crowcroft. 1996. Quality-of-service routing for supporting multimedia applications. *IEEE Journal on Selected Areas in Communications*, vol. 14, no. 7, pp. 1228–1234.

24. Reeves, D. S., and H. F. Salama. 2000. Distributed algorithm for delay-constrained unicast routing. *IEEE/ACM Transactions on Networking*, vol. 8, no. 2, pp. 239–250.

25. Varga, A. 2003. OMNeT++—Discrete event simulator. https://omnetpp.org/ (accessed 2011).

26. Eiza, M. H., Q. Ni, T. Owens, and G. Min. 2013. Investigation of routing reliability of vehicular ad hoc networks. *EURASIP Journal on Wireless Communications and Networking*, vol. 2013, no. 179, pp. 1–15.

27. Monteiro, J. 2008. The use of evolving graph combinatorial model in routing protocols for dynamic networks. In *XV Concurso Latinoamericano de Tesis de Maestrìa*, Santa Fe, Argentina, pp. 1–17.

28. Vegni, A. M., and V. Loscri. 2015. A survey of vehicular social networks. *IEEE Communications Surveys & Tutorials*, vol. 17, no. 4, pp. 2397–2419.

Chapter 4

Revealing the Role of Structural Transitivity in Building the Sustainable Community-Aware Vehicular Social Networks

Syed Fakhar Abbas, William Liu, Quan Bai, Adnan Al-Anbuky, and Aminu Bello Usman

School of Engineering, Computer and Mathematical Sciences, Auckland University of Technology, Auckland, New Zealand

Contents

4.1 Introduction

Vehicular communication networks are a type of network in which vehicles and roadside units are the communicating nodes, providing each other with information such as safety warnings and traffic information. As a cooperative approach, vehicular communications can be more effective in avoiding accidents and traffic congestion, rather than each vehicle trying to solve these problems individually. Therefore, intelligent transportation systems (ITS) have been developed to address the challenges of the safety, security, and efficacy of the current transportation systems. The field of inter vehicular communications (IVC) includes both vehicle-to-vehicle (V2V) and vehicle-to-infrastructures (V2I) communication and is also known as the vehicular ad hoc network (VANET). The VANET is recognized as an important component of ITS in various national plans [1]. Direct communication between vehicles may be set up by means of mobile ad hoc networks (MANETs), which do not depend on the altered foundation. Research on MANETs covers the application prerequisites and correspondence conventions for everything from sensor systems to handheld PCs and vehicular frameworks [2].

ITS are propelled applications plan to give creative administrations identified with distinctive methods of transport and activity administration, through vehicular correspondence, to enhance human safety and provide more safety to drivers. Autos furnished with remote gadgets can trade activity and street security data with adjacent autos, or roadside units. Vehicular networks have turned into a prominent research subject in the most recent years, because of the vital applications that can be acknowledged in such a situation. In [3], the authors have separated such applications into two important classifications: Internet connectivity and peer-to-peer applications.

In vehicular systems, messages between vehicles can be utilized to distinguish distinctive levels of car influxes [4], and in this way, movement speed can be decreased with the help of V2V correspondence [5]. As of late, the creators in [6] displayed how IVC can decrease the number of auxiliary crashes brought about by an accident, through a scattering of caution messages. A later overview of other applications and use cases can be found in [7]; the authors arranged them into three classes [8]: management and traffic controlling applications, road safety applications, and commercial applications. The authors in [2] explain more about safety and entertainment applications.

Nonetheless, VANETs show bipolar conduct contingent upon system topology: completely joined topology with high movement volume joined topology when movement volume is low [9]. Along these lines, one can recognize two diverse classifications of vehicular systems: VANETs, which are presented above, and vehicular delay-tolerant networks (VDTNs), which are vehicular systems in delayed movement, and where delay-tolerant network (DTN) conventions can be connected. To ensure the possibility of numerous applications through vehicular systems, it is basic to plan conventions that can overcome major issues that emerge from vehicular situations. Besides, Internet conventions do not function admirably for some situations [10], because of some principal suppositions incorporated with the Internet's structural engineering, for example, the presence of a conclusion to end way in the middle of source and destination for the span of a correspondence session, short end-to-end round-outing postponement time [11], and the recognition that bundle exchanging is the most proper deliberation for interoperability and execution.

The high speed and rate of nodes in vehicular situations are in charge of an exceptional element system topology that is unique in relation to the customary idea of the Internet. These nodes can show short contract terms, or move in an eccentric way [12]. The connections may be fleeting, with high connection mistake rates, and the nonattendance of a conclusion to end way from the source to the destination. Thus, arrangements in such situations can be apportioned, because of the vast separations included and the variable node densities and meager movement, bringing about discontinuities along the way from the source to the destination [13]. Numerous traditional steering conventions were intended for VANETs on account of a completely interconnected environment, planning to set up end-to-end availability among system vehicles [14]. On the other hand, these conventions cannot be utilized when the activity calms down. End-to-end between vehicles cannot be set up any more [15]. Consequently, this classification of steering conventions neglects to convey information in inadequate activity, apportioned systems, and sharp vehicular systems.

While trying to address this issue, vehicular systems might convey information utilizing the store-carry-and-forward (SCF) worldview of DTNs [10], as opposed to a basic convey-and-forward technique. Hence, long and variable-length messages, called groups, can be sharply steered toward the destinations through discontinuous associations, accepting that end-to-end system way is not, as a matter of course, presently accessible, but instead that such a way exists after some time. In this way, DTNs in the vehicular environment are called VDTNs [14]. In VDTNs, communications between vehicles show up with no past information [15], and in this way the difficulties that DTNs need to overcome have prompted critical exploration concentration. The thought of bunch association has been utilized for MANETs in a number of issues, for example, directing, security, quality of service (QoS) [16,17]. However, the qualities of VANETs, for example, speed, rapidly changing the movement of vehicles, and topology, are a major difference that why MANET routing protocols are not suitable for VANETs.

In this chapter, we evaluate how the transitivity and social community-aware approach affect the information sharing and transmission between vehicles. The rest of chapter is organized as follows. Section 4.2 reviews the related work in structural transitivity and social community–aware routing approaches. The structural transitivity and its implications are presented in Section 4.3. In Section 4.4, extensive simulation studies have been conducted and results analysis is presented. Finally, the conclusions and layout of future work are drawn in Section 4.5.

4.2 Related Work

The VANET is one of the emerging wireless communication network areas. It is the upcoming area MANET where vehicles work as portable nodes within the network. Comforting passengers and increasing road users are some of the basic targets of VANET. Its communications take place through wireless links that are mounted on each and every node, that is, the vehicle [18].

The examination of informal organizations was started by Milgram in the 1960s [19]. Milgram presented the thought of little world marvel, which demonstrates that any pair of individuals in the world can interface with one another through little successions of connections (commonly five or six), and from there on, numerous works reaffirmed it [20–22].

A few works have been accounted for in VANETs managing versatile examples of vehicles and the development of bunches. The investigation of VANET qualities and distinctive versatile examples of vehicles is exhibited in [23]. An auto taking after a model utilizing a neural system approach for mapping observations to activities is created in [24]. The model has a definition to the craved separating model that does not consider response time and endeavor to clarify the behavioral parts of the taking-after auto. An entombed vehicle data dispersal convention called the got message-subordinate convention (RMDP) that engenders the former activity data to the accompanying vehicles is discussed about in [25]. The RMDP convention self-rulingly changes the dispersal interim, relying upon the quantity of gathering messages and recognized gathering blunders, keeping in mind the end goal to stay away from message crashes among vehicles. Utilizing RMDP, numerous vehicles can get going before movement of data inside of short periods under light and overwhelming movement conditions.

A social bunch–based point-to-point structure that gauges similitude and association conditions among portable companions to give proficient asset revelation and recovery is proposed in [26]. Portable peer's inclination and its availability, for example, the lifetime and transfer speed of a remote connection, are considered. A portability-based stable bunching plan for VANETs that uses the proclivity proliferation calculation in a conveyed way is displayed in [27]. Vehicular versatility amid bunch arrangement is considered in the liking proliferation calculation. Every vehicle in the system transmits the obligation, as well accessibility messages, to its neighbors and afterward makes a free choice on bunching. A restriction procedure that exploits the developing VANET environment is considered in [28]. Correspondence among vehicles is used to figure out the relative vehicle area, the reconciliation of which with movement data furthermore, GPS area evaluation leads to precise vehicle restriction. Bunching of vehicle directions obtained by a mechanized vision framework is proposed in [29]. A trajectory similarity measure is performed taking into account the handoff separation, with changes to enhance its power and record the actuality in which directions are requested accumulations of focuses.

A reference point–based grouping calculation for dragging out the bunch lifetime in VANETs is discussed in [30]. A dispute strategy is utilized to abstain from activating regular redesigns at the point where two bunch heads experience one another for a brief timeframe. Three uninvolved bunching (PC)-based strategies to decide which suitable vehicles will be the principle members in group structure development are examined in [31]. PCs consider various measurements, for example, vehicle speed, join quality, and connection maintainability. A bunching calculation and a various leveled directing convention that cooperate to accomplish the system soundness are exhibited in [32]. Three measurements for measuring system soundness are exhibited: (1) the group lifetime, (2) the intercluster connection lifetime, and (3) the end-to-end way lifetime.

Every node declares itself a bunch head by putting its own location and ID in the show guides. Subsequent to accepting guides from neighbors, a vehicle has complete learning of its present neighbors and determines whether to change its present bunch status. The status change is bidirectional, either from bunch head to part vehicle or from part vehicle to group head. The choice to change status is done relying on three main considerations: vehicle ID, current course, and administration term.

Vehicular clustering in view of the weighted clustering algorithm (VWCA) is exhibited in [34]. VWCA takes neighbors into account in view of the element transmission run, course of vehicles, entropy, and doubt esteem parameters. A versatile assignment of transmission range strategy is examined to adaptively alter the transmission range among the vehicles. A decentralized and versatile approach for data spread in VANETs is examined in [35]. The effectiveness of versatile methodologies in data spread compared with factually based methodologies is concentrated on in this work.

4.2.1 Vehicular Social Networks

The idea of social cars emerges from the supposition that every driver can transfer information to different neighbors in light of normal hobbies; for example, Ford's idea auto Evos can straight-forwardly shape an informal community with drivers companions [36]. Beginning from essential components of VANETs, our point is to show how human social conduct can change the way autos are driven in the next couple of years. Today, person-to-person communication is a reality, and presenting social angles in VANETs permits vehicles to convey, as well as select, comparative neighboring in light of social measurements.

4.2.1.1 Data Dissemination in VSN

Recent accomplishments in the setting of the information spread approach in vehicular social networks (VSNs) are most importantly considered in [37], where the creators recognize three principal classes in view of (1) data preparation, (2) content conveyance, and (3) execution. Fundamentally, the primary thought for the configuration of substance scattering conventions and steering calculations in VSNs misuses social properties, as well as the portability conduct of people and vehicles. Xia et al. [38] present artificial bee colony–enlivened interest-based forwarding (BEEINFO), a steering system that characterizes groups into indicated classifications, on the premise of individual hobbies. The general thought of BEEINFO is that versatile vehicles see and record data (e.g., vehicular densities) of passing groups, comparative to how honey bees fly from one bloom to another. The similarity data shows the quantity of vehicles having a place with a group: the higher the similarity, the more vehicles the group has. This data gives a rule to better select next-jump forwarders.

Essentially, a VSN is included two key parts: (1) a vehicular specially appointed system that speaks to the physical layer, and (2) an informal community system running on top of such a physical vehicular system. Subsequently, a VSN needs solid participation between social angles and physical system operational instruments. In [39], Fei et al. consider a VSN plan in an urban vehicular situation. The IEEE 802.16j innovation is utilized to empower vehicular interchanges and some methodologies concentrated on the dispersal schedule of the 802.16 standard [40] and [41] to enhance the use of transmission capacity assets. A relay station (RS) is a transport that conveys various clients, while a roadside base station (BS) serves different moving RSs inside of the scope; after that, a RS might promote numerous administration subscriber stations (SSs). The BSs are associated with the Internet by means of Web access gateways. The VSN supplier is likewise associated with the Internet and gives an online gateway to intrigued clients to enlist and utilize its interpersonal interaction administrations. With a specific end goal to better comprehend the conduct of social-based vehicular systems, we use the principle devices of social network analysis (SNA) [42]. SNA considers social connections as far as vehicles (i.e., people) and ties (i.e., connections among vehicles), and recognizes vital parts in an informal community, for example, the centrality measurements that are utilized to signify how "vital" a vehicle is inside a system. In fact, an informal community is comprised of clients; social ties, on the other hand, are connections among clients and basic hobbies. Each of the three parts might affect the social impact of clients [43]. As known, a VANET is a continually developing system, with motion that changes after some time. Therefore, one of the primary elements to look at is the system network after some time, accepting that vehicles can construct astute moving network joins. SNA can be used to check the activity advancement during the day, to understand the human schedules, comparable directions, and surge times.

To completely comprehend the motion of a VANET, numerous scientists [44–47] have concentrated on the conduct of vehicles using a method for the substantial size of vehicle directions over genuine street systems. In [44], Papadimitriou et al. studied the structure and development of a VANET by utilizing practical vehicular follows from the city of Zurich. In particular, they focused on a 5 km × 5 km street range, covering the focal point of Zurich, and containing around 2 × 105 unmistakable vehicle directions in a run-of-the-mill morning rush hour. In such a situation, the dissemination of the centrality measurements is not influenced by the correspondence range, but rather, it relies on the variety in movement conditions, that is, speed and relative positions of the vehicles. Thus, centrality is not an ancient rarity of the correspondence range, but a component of the conduct of the vehicles, that is, street system and driver expectations.

In [45], Cunha et al. present a numerical examination of genuine but reasonable information sets that portray the versatility of vehicles, from a social point of view. The creators show that the vehicular situation influences the vehicles' velocity, then affecting the experience proportion. Likewise, the way of vehicles influences the amiability viewpoints in vehicular situations; for example, taxis cross the entire city in arbitrary directions, without a settled time, as well as, outing span, while transports travel the same courses under an altered timetable, and regular individuals utilize their vehicles to go in preordained directions as per their schedules. At last, in [46] the appraisal of a basic show information spread convention in VANETs is given. The configuration of an ideal arrangement of handoff vehicles, upgrading framework execution, has been examined for distinctive movement situations (i.e., parkway, country, and urban) in the city of Rome, for (1) between vehicular correspondences, and (2) accessibility of the altered system foundation for V2I correspondences. An itemized depiction of information dispersal conventions for VANETs is introduced in [48].

Utilizing past works, the network conduct in VANETs can improved the key components, for example, (1) vehicles' versatility design, (2) transmission range, (3) the presence of a system base, and (4) market entrance [49]. The driver's conduct produces awesome impacts in vehicular versatility; for example, individuals have a tendency to go to the same spots, at the same time of day, in the same directions. At that point, vehicles experience others vehicles, go in the same avenues, and endures the same movement conditions. Every one of these components recommends (1) the investigation of the vehicular versatility under a social point of view, and (2) application of social ideas to enhance the administration and availability in VANETs.

4.2.1.2 Social Features in VSN

As officially expressed, social qualities and human conduct to a great extent sway vehicular systems, and along these lines, VSNs has emerged [50]. The impact of human variables, that is, generally human versatility and childish and client inclinations, to a great extent sways vehicular availability. In [51], Cunha et al. present the portrayal and assessment of a reasonable vehicular follow to consider the vehicles' versatility in connection to social practices. Through numerical investigation, the creators distinguish impossible-to-miss social attributes in vehicular systems, and how the utilization of these measurements can enhance the system execution of correspondence conventions, as well as benefits. To be sure, human variables are included in vehicular systems because of security-related applications, as well as for non-security-related applications, that is, amusement. From the nature of vehicular specially appointed systems, activity examples can give social connections. As an occurrence, in overwhelming movement situations (e.g., amid morning rush hours), the vehicular speed is high and the movement example is generally static. Such situations turn into a mainstream social spot for vehicles to unite with each other and offer data (e.g., activity data and climate news).

In [55–56], the issue of stable vehicle grouping is examined, with a specific end goal to restrict the telecast storm issue. Without a doubt, due to the quickly changing system topology, vehicle groups are manufactured powerfully, and information bundles can be sent at various times. As an answer, Maglaras and Katsaros [52] build up a sociological pattern clustering (SPC) and route stability clustering (RSC) calculation, misusing the social conduct of vehicles, that is, their propensity to have the same or comparable courses. Portability models for VSNs are influenced by (1) the human portability model, (2) the human narrow-minded status, and (3) human inclinations. In VSNs, vehicles are driven by individuals with their own abilities and driving style (i.e., smooth deceleration, increasing speed, and astute driving patterns). For instance, drivers choose the most direct way to a destination, rather than going the longest way.

Other portability models take after aggregate human conduct (i.e., group). In the group-based portability model, it is accepted that there exist a few places with high social appeal (e.g., eateries, shopping centers, and theaters). At last, portability in VSNs is additionally influenced by a period variation model; for example, a vehicle moves toward a given spot at a given time of a day (e.g., individuals go to the workplace in the morning, and back home at night, while on Sunday, individuals want to unwind at home, and afterward, activity is low in the urban region. A case of a portability model for social-based vehicular systems is introduced by Lu et al. in [55,56]. The creators examine VANETs as far as social-nearness highlights, following numerous vehicular situations in the nearness-related applications, for example, security message scattering and restricted social substance sharing. Lu et al. present a portability model called Restricted Mobility Region with Social Spot, where the urban region is a versatile framework with an arrangement of social spots so that the portability locale of every vehicle is confined and connected with a settled social spot.

4.2.2 Social-Aware Community Routing Approach

As VANETs have been utilized as a part of different applications whose extreme objective is to give well-being and solace to the travelers sitting in the vehicles, there is a necessity of improved arrangements for grouping in VANETs. Be that as it may, because of a huge number of nodes, as well as the absence of switches, a level system plan may bring about genuine adaptability, as well as shrouded terminal issues. A conceivable answer for the above issues is the utilization of a productive grouping calculation. With respect to efficient correspondence among the vehicles out and about, dedicated short-range communication (DSRC) is utilized, so it would be a logical thought to partition the vehicles into groups so that vehicles inside of the same group may convey utilizing DSRC benchmarks. These truths inspire us to classify different grouping methods in VANETs based upon predefined criteria. Be that as it may, there are a number of difficulties that need all-around outlined answers for grouping of vehicles.

A portion of the difficulties is high portability of the vehicles, scanty network in a few areas, and security. Because of the extensive number of parameters that have been considered in the distinctive grouping, it was hard to think of some as the standard for assessment of inspected conventions. To oblige this assorted qualities, every one of the parameters is broken down and blended into eight standard classes in this chapter. These eight parameters have been extensively ordered into vehicle speed and vehicle velocity, which describe the effectiveness of the grouping convention; bunch solidness; group flow; group joining; group associate time; transmission productivity; and transmission overhead. This categorization will assist us with providing a near examination of all the evaluated grouping conventions. A portion of the above-depicted parameters is outlined below. Vehicle density, which is a standout among the most critical parameters, characterizes the normal number of vehicles in vehicles per kilometer or vehicles per path.

For urban situations a high estimation of vehicle speed is considered, contrasted with expressways. Vehicle rate is the scope of rates considered for reproduction by a specific convention in meters per second or kilometers per hour. A rate range that fluctuates practically shows better flexibility. Transmission efficiency is portrayed as a normal number of messages or parcels that are transmitted or received by a group part in a period. High transmission productivity demonstrates that a bunching plan is more compelling in information spread. Transmission overhead is the normal correspondence or control overhead required by a bunching plan for bunch development and upkeep in number of packets.

A grouping plan that has lower transmission overhead is wanted. Group solidness is the normal lifetime of a group. A high estimation of group solidness demonstrates a superior bunching convention. The parameter group motion portrays the normal number of status changes per vehicle characterized regarding a normal number of group changes or bunch head changes in the aggregate number of vehicles. A low estimation of group motion is more suitable. Group interface time alludes to the length of time that a vehicle stays associated with a solitary group. A high estimation of group interface time demonstrates the higher suitability of a convention. Group meeting is in reference to the length of time required for each one of the nodes to join a group at the start of a bunching plan. The suitability of a grouping plan for VANETs is more when it displays low group merging.

These days, a few car makers are looking forward to achieving the objectives imagined by the Vision 2020 activity plan.[1] In particular, in 2012 the European Commission tabled the CARS 2020 action plan, fortifying this present industry's intensity and manageability heading toward 2020.

4.2.3 Routing Protocols

The most effortless approach to convey a message to its destination node in a delay-tolerant system with a high likelihood of achievement and negligible postponement is to utilize spread-based steering (also known as a type of controlled flooding). The maker of a message starts sending the information to everyone of its neighbors. Neighbors are characterized as all vehicles inside of correspondence reach. A neighbor who did not get the message some time recently will forward it to all its neighbors. This procedure will continue until all nodes that are a piece of the system are educated. Spread-based steering has a tendency to minimize the transmission time of messages in deft systems; however, it requires a great deal of data transfer capacity, which can prompt system blockage. To moderate this downside, the range of a message can be constrained to an uncritical quality. There are a few approaches to doing this; for example, the quantity of jumps that a message is permitted to pass can be constrained or a replication component can be characterized. The replication element impacts the quantity of message duplicates that are simultaneously permitted to be available in the system.

A surely understood illustration of such spread-based directing conventions is Epidemic Routing [57]. It is a relationship to the spreading of scourges by pairwise contact between individuals. In an artful system, a vehicle is contaminated with an infection (the node stores the message locally for conveyance to other nodes) when it creates the message itself or the message gets sent from another vehicle in the system. As soon as a tainted vehicle conveys the message to its destination vehicle, the tainted vehicle is amended. Starting there in time, it is resistant to the infection and does not acknowledge this message once more. As of now demonstrated, this methodology is not common sense due to the exponentially developing correspondence overhead.

Like Epidemic Routing, the Probabilistic Routing Convention utilizing History of Encounters and Transitivity (Prophet) [58] is a variation of the flooding calculation. The contrast is that messages are just handed off to those neighbors, which are thought to bring a higher likelihood of conveying the message to its destination than that vehicle which presently holds the message. These forward-looking suspicions depend on a background marked by vehicles, which have been experienced before.

4.3 Structural Transitivity and Its Implications in V2V Communication

4.3.1 Transitivity

Transitivity is one of the important measures for the presence of 3-clusters and a critically driven factor of cohesiveness in the network and for efficient communication between devices. The study in [59] describes transitivity as a critically driven component of trust and notoriety in the complex system of relationships between peers in a network. Additionally, transitivity has been utilized to actualize a well-known group discovering system and the making of connections in an interpersonal organization, taking into account the idea of the presence of triads, which are more likely to appear between symmetry nodes in the network. This concept has been used to implement a popular community finding method and the creation of links in a social network based on sharing a common node.

4.3.2 Structural Transitivity Model

In particular, we are more interested in the proportion of transitive 3-clusters (which express the degree of balance in the network), which most social science and network science theorists argue are the "equilibrium" or natural state in which 3-cluster relationships efficiently tends to a normative relationship between the actors. In reflecting the social network transitivity between vehicular network devices, we consider a network to be transitive if it satisfies the following requirements:

$$V1, V2, V3: V1 \ R \ V2 \wedge V2 \ R \ V3 \rightarrow V1 \ R \ V3$$

That is, if there is a tie between Vehicles $V1$ and $V2$, and there is also a tie between Vehicles $V2$ and $V3$, then this will result in the tie between $V1$ and $V3$, as illustrated below:

We further employ the concept of clustering coefficient matrix to quantify the transitivity between the peers of vehicles in the network. We use a local clustering coefficient as a measure of motives clustered in the network. In the study of network and social science, a clustering coefficient is used to measure the degree of transitivity and interdependency between actors of local relations in the network and the level of sharing a common neighboring node between actors of the cluster. We define the local clustering coefficient as follows:

$$Local \ clustering \ coefficient \ C = \left(\frac{|Paths \ of \ length \ 2 \ that \ have \ the \ third \ edge|}{|Paths \ of \ length \ 2|} \right)$$

The local clustering coefficient measures transitivity at the node level $C(vi)$,

$$C(vi) = \left(\frac{No. \ of \ pairs \ of \ neighbors \ of \ vi \ that \ are \ connected \ to \ vi \ to \ form \ a \ triangle}{Number \ of \ pairs \ of \ vi's \ neighbors} \right)$$

Let $G(V, E)$ be an undirected graph without loops on the vertex set $V = \{V = \{vi, vj, ..., vn\}\}$, where n is the order of the graph and a k subset of V is a subset of V containing k vertices. If we denote $\binom{V}{k}$ as the class of all k subsets of V, we have for E the edge set of G such that $E \leq \binom{V}{2}$; the elements of Y and the corresponding $n \times n$ adjacency matrix of G are given by

$$y_{ij} = y_{ji} = \begin{cases} 1 & \text{if } \{i,j\} \in E \\ 0 & \text{otherwise} \end{cases} \tag{4.1}$$

$$r = |E| = \sum_{i=1}^{n-1} \sum_{j=i+1}^{n} y_{ij}. \tag{4.2}$$

Since the transitivity and triads are interrelated, it will be almost inevitable for us to deal with triad counts. Triads can have zero to three edges, that is, be of size 0–3. We may use the adjacency matrix Y in order to obtain the count of induced triads of size ℓ as

$$t_\ell = \left| \left\{ \{i,j,k\} \in \binom{V}{3} : y_{ij} + y_{ik} + y_{jk} = \ell \right\} \right|. \tag{4.3}$$

Δ_ℓ denotes the proportion of complete subgraphs of order ℓ out of all possible subsets of order ℓ; thus, for $\ell > 2$,

$$\Delta_\ell = \frac{\left| \left\{ C \in \binom{V}{\ell} : \binom{C}{2} \subseteq E \right\} \right|}{\left| \binom{V}{\ell} \right|} = \left(\sum_{C \in \binom{V}{\ell}} \prod_{\{i,j\} \in \binom{C}{2}} y_{ij} \right) \binom{n}{\ell}. \tag{4.4}$$

If we let $N = \binom{n}{2}$ denote the number of all positions in the graphs where there could be edges, we realize that for $\ell = 2$,

$$\Delta_2 = \left(\sum_{i=1}^{n-1} \sum_{j=i+1}^{n} y_{ij} \right) \cdot \binom{n}{2} = \frac{r}{N}. \tag{4.5}$$

That is, Δ_2 is the graph density for $\ell = 3$; thus, we have that

$$\Delta_3 = \left(\sum_{i=1}^{n-2} \sum_{j=i+1}^{n-1} \sum_{k=j+1}^{n} y_{ij} y_{ik} y_{jk} \right) \cdot \binom{n}{3} = t_3 = \binom{n}{3}. \tag{4.6}$$

In other words, Δ_3 is the proportion of transitive triads out of all triads. This proportion can be used as a transitivity index. The transitivity index of a graph is the ratio of the number of triangles to the number of all possible links in the graph, a real number between 0 and 1. If the distance between nodes is small, then their transitivity values are higher, which means they have the shortest path and better way to transmit the data to other nodes.

4.3.3 Carbon Emission Model

Transport emissions are among the highest environmental pollution sources in cities [60]. Vehicles emissions contain a number of harmful substances. These include fine particulate matter, carbon dioxide (CO_2), carbon monoxide (CO), hydrocarbons (HCs), nitrogen dioxide, and nitrogen monoxide. Refining the mobility of transportation and bettering the living environment are two essential problems that need to be carefully handled in urban traffic system [61]. In general, those factors can be ascribed to a road traffic accident situation (icy, closed), bad driving behavior, and traffic signs and lights.

From the archives of the literature, there are two most popular methods uses for carbon emission estimation. The first method is to estimate the energy consumption at different vehicular speeds in miles per hour. The second method is a comprehensive modal emissions model (CMEM) [70]. The CMEM has three main functions: providing detailed fuel consumption, estimating localized emissions, and accounting for road grade effects. These three functions are helpful for estimating the carbon emission of each vehicle [60,62]. The first method evaluates CO_2 emissions based on average speed of a trip or trip segment. According to the research, the air pollutant released by burned fuel will increase during acceleration [26]. We can use the Equation 4.7 to compute the carbon emission.

$$ln(y) = b0 + b1 \cdot x + b2 \cdot x2 + b3 \cdot x3 + b4 \cdot x4 \tag{4.7}$$

where y stands for CO_2 emission in grams per mile and x stands for the average travel speed in mile per hour. The concrete number that every letter stands for is in [60,63]. With the total mass of the system, the engine, the speed, and the acceleration of the calculation, we can produce analog data. This data will be treated as an estimate of the emissions: The CO_2, CO, HCs, nitrous oxide (NO_x), and so on. By calculating the acceleration and deceleration of the vehicle emissions, we can estimate CO_2 emissions. The tractive power requirement at a vehicle's wheels P_{tract} is calculated using the following polynomial:

$$P_{tract} = Av + Bv^2 + Cv^3 + Mav + mgv\sin\theta. \tag{4.8}$$

Based on the tractive power requirement, the gas consumption can be estimated and, consequently, the tailpipe emissions of CO_2 can be calculated according to the following polynomial:

$$TP_{CO_2} = \begin{cases} \alpha + \beta v + \delta v^3 + \acute{C}av & \text{if } P_{tract} > 0 \\ \alpha' & \text{else} \end{cases} \tag{4.9}$$

Because road grade is not currently modeled, P_{tract} calculations assume planar roads, and hence, $\vartheta = 0$ [63]. Values of α to $\acute{C}$, A to C, and M were fitted to match what authors termed a category 9 vehicle,[4] for example, a 1994 Dodge Spirit. The used values (Table 4.1) result in an error in CO_2 emission calculations of approximately 2.2%.

Also, the numerical value of carbon emission in 0–20 mph is highest at 0–90 mph (i.e., relationship between average speed and carbon emission). Meanwhile, the numerical value of carbon

Table 4.1 Emission Factors for a Category 9 Vehicle

Factor	Value	Unit
v Vehicle speed		$m\ s^{-1}$
a Vehicle acceleration		$m\ s^{-2}$
A Rolling resistance	0.1326	$kW\ m^{-1}\ s$
B Speed-correction to rolling resistance	2.7384×10^{-3}	$kW\ m^{-2}\ s^2$
C Air drag resistance	1.0843×10^{-3}	$kW\ m^{-3}\ s^3$
M Vehicle mass	1.3250×10^{-3}	Kg
g Gravitational constant	9.81	$m\ s^{-3}$
ϑ Road grade	0	Degrees
α Index of a lane		
θ Slope of a road segment		
$\emptyset$ Index of a road segment		
γ Length of a lane		
λ Index of a vehicle type		

emission in 0–20 mph will be on the rise. Actually, the vehicle will stay at low speed in traffic congestion, which means the vehicle will release more air pollutant, including carbon emission [64]. After traffic congestion ends, the vehicle will accelerate to get up to the normal speed. In acceleration, it will also release more carbon pollutants. In view of this, the relationship between traffic congestion and carbon emission is that traffic congestion will cause more carbon emission [64]. Moreover, we can equally employ the use of the Equation 4.10 to compute the CO_2 emission:

$$\text{Emissions}_{CO_2} = \left(\frac{\text{Liters}}{\text{Kilometer}} \right) \cdot \left(\frac{\text{Mass } CO_2}{\text{Liter}} \right) \cdot (\text{Kilometers traveled}) \qquad (4.10)$$

Liters per kilometer is a vehicle technology-related parameter that is used to describe the fuel energy economy of a vehicle, while mass CO_2 per liter is the fuel-related parameter for estimating the amount of CO_2 emission produced by each unit of fuel, and the kilometers traveled is the traffic activity parameter representing the travel distance of total vehicles. Technically, the shortest path between the two vehicles determines the sum of the weights of the ties between the vehicles. We therefore consider the definition of the shortest path metric of the undirected, unweighted network as follows:

$$Spath_{V1,V2} = \sum_{V1,V2 \in G} \frac{d(V1,V2)}{n(n-1)} = \sum_{i,j \in E} \frac{d(yij)}{n(n-1)} \qquad (4.11)$$

where $V1$ and $V2$ are the two fairs of vehicles and n is the number of possible vehicles between $V1$ and $V2$. From Equation 4.11 (average shortest path between the vehicles) and 4.6 (proportion of transitive triads out of all triads), we can deduce that the two equations are directly proportional, meaning that when the average structural transitivity coefficient of vehicles is increasing, the average shortest path is also increasing, which by extension will result in better and efficient data communication between the vehicles. In addition, when the vehicles have higher transitivity, they can easily establish efficient communication with other vehicles, so that they can easily share information with those other vehicles in the case of an accident, road blockage, or so forth, which leads to less carbon dioxide emission, since vehicles can use alternate paths.

4.4 Case Studies

In this section, we define how the information sharing in structural transitivity using social community awareness will perform compared with information sharing without social effects. Afterward, we present a scenario where our proposed algorithm can be adopted, and we explain and justify the simulation parameters we have used. Finally, we show and explain the results obtained through simulations.

4.4.1 Network Scenario

The simulation has been performed using the Opportunistic Network Environment (ONE) simulator [65]. ONE is an operator-based discrete-occasion reproduction motor. At every reproduction step, the motor upgrades various modules that execute the fundamental reproduction capacities. The primary elements of the ONE test system are the demonstrating vehicle development, between-vehical contacts, and directing and taking care of messages. Results gathering and examination are done through perception, reports, and posthandling apparatuses. The ONE simulator provides support for the map-based movement, based on street graphs that are given in WKT format. However, the manual creation of such maps is time-consuming and error-prone, and purchasing high-quality map data is very expensive. The OpenStreetMap project is community-driven effort to create street maps of the entire world that are free of use. It is based on an idea similar to that of the well-known Wikipedia project: contributions from users. OpenStreetMap allows the export of street data from the website in OpenStreetMap XML (OSM) [66] format. For running simulations for different street maps, the osm2wkt [67] program has been used; osm2wkt can convert OSM into WKT for running a simulation in the ONE simulator based on OpenStreetMap. The scenario we have used is based on Auckland Queen Street (Figure 4.1).

We got a map from OpenStreetMap and then converted the OSM map into WKT for use in the ONE simulator. The scenario consists of almost 100 nodes in which we have created communities of nodes. So, each community includes 10 nodes. Node movement is actualized by movement models. These are either engineered models or existing development follows. The network between the vehicles depends on their area, correspondence range, and bit rate. The routing for direct capacity is actualized modules that choose which messages to forward over existing contacts. At last, the messages themselves are created through occasion generators. The messages are dependably unicast, having a solitary source and destination host inside the recreation world. Recreation results are gathered fundamentally through reports produced by report modules amid the recreation run. Report modules receive occasions (e.g., message or network occasions) from the recreation motor and create results in light of them. The outcomes produced may be logs of occasions that are then further handled by the outside postpreparing apparatuses, or they may be total insights ascertained in the test system. Optionally, the graphical user interface (GUI) shows a perception of the reproduction state appearing in the areas, dynamic contacts, and messages conveyed by the nodes.

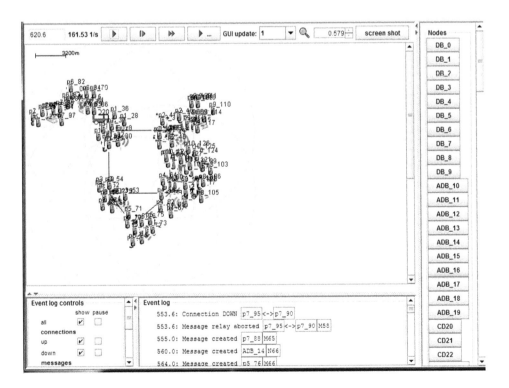

Figure 4.1   Simulation scenario.

The focus of the ONE is on the evaluation of a protocol's behavior and functionality, which was developed on the basis of the SCF principle. A complete implementation of the lower network levels has been omitted. Instead, simplified assumptions about data rates, communication ranges, and so forth, serve as parameters for the simulations.

4.4.2 Simulation Studies

We have used the customized setting file. The simulation area is 1500 m × 1500 m; also, we have used the OpenStreetMap export in OSM format and then converted the OSM file into WKT format which is supported by the ONE simulator. The movement of the nodes is a random walk with random speed between 10 and 14 m s^{-1}. The simulation parameters are defined in Table 4.2, where routing protocols mention which they will use and so forth.

4.4.3 Results Analysis

4.4.3.1 Message Delivery Ratio

In this section, the different routing protocols Epidemic Routing and Prophet are used with and without social-aware communities. Figure 4.2 compares the routing message delivery ratio with and without communities achieved by each routing protocol during the performed simulations. It shows worse performance when disseminating information without social communities in both routing protocols. Only 40% and 35% message delivery ratios are achieved using the Epidemic and Prophet routing protocols, respectively, without social-aware communities, but when we simulate using same routing protocols with social-aware communities, the message delivery ratios are much higher. The message delivery ratio is higher in social-aware communities using transitivity because data broadcast sharing in small social-aware communities is faster as than information sharing in a higher number of nodes; also, transitivity builds the shortest and quickest path between nodes, which result, in a higher data message delivery ratio.

4.4.3.2 Latency

Data latency is the time it takes for data packets to be stored or retrieved. Figure 4.3 shows the latency of social-aware community and noncommunity social-aware routing. From Figure 4.3, we can see that the social-aware community has lower data latency because routing information is easily passed and shared in the social system. Obviously, a large communication range facilitates increases in the latency rate. We then compare the latency of social-aware communities with that

Table 4.2 Simulation Parameters

Parameters	Values
Simulator	Opportunistic Network Environment (ONE)
Routing protocols	Epidemic and Prophet
Social add-on for communities	dLife and dLifeComm
Simulation time	1800 s (0.5 h)
No. of nodes	100
No. of communities	10
No. of nodes in each community	10
Node interface	13–27 Mbps

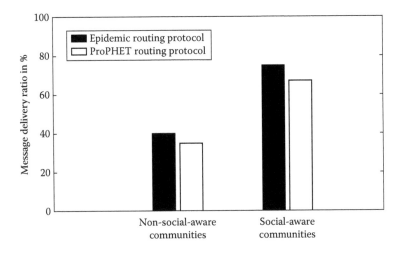

Figure 4.2 Comparison of message delivery ratio of communities and noncommunities.

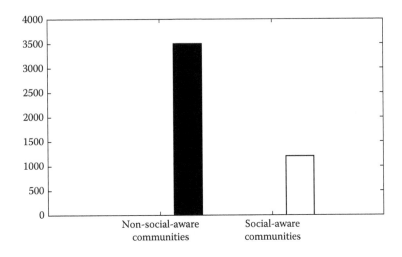

Figure 4.3 Latency of social-aware communities and non-social-aware communities.

of non-social-aware communities. The social-aware communities perform well because they have a transitivity effect, which builds the quickest and shortest path between vehicles, resulting in lower latency.

4.4.3.3 Carbon Emission

As we can easily see from Figure 4.4, less carbon emission is generated by the vehicles that have a social-aware community social network feature. On the other hand, the scenario in which the social-aware feature is not used within the vehicle produces more carbon dioxide emission. Thus, we can easily that social-aware features help vehicles to better communicate and produce less CO_2. As we clearly analyzed from Figure 4.4, social-aware communities have less carbon dioxide emission

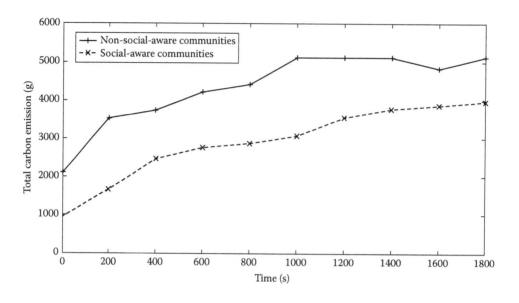

Figure 4.4 Carbon emission of social-aware communities and non-social-aware communities.

than non-social-aware communities because transitivity builds the shortest path between vehicles for communication in the case of any road blockage, accident, and so forth. Vehicles use an alternate path, which results in less CO_2 emission.

4.4.3.4 Packet Dropped

The data dropped ratio is higher in non-social-aware communities because social-aware communities divide the vehicles into clusters and better transfer the data information, and data accurately delivered to the vehicles belongs to clusters, therefore, rather than sending data to all vehicles, these communities only send data to cluster nodes, which results in less data dropped, as well as transitivity to build the fastest paths between nodes, which also results in less data dropped. As we have clear seen from the Figure 4.5, social-aware communities have less data dropped than non-social-aware communities because the transitivity vehicles form a quick path between vehicles.

4.5 Conclusion and Future Work

This chapter has presented a social-aware packet forwarding and routing framework by considering the structural transitivity of the underlying vehicle network topology. The nearby association, that is, the transitivity between mobile vehicles and drivers, social connections, pulls in specialists to investigate the capability of bringing social properties to ITS. In addition, we have discussed the basic concepts behind this new terminology and presented the first survey of the state of the art on how the underlying topological structure can impact the overlay network communications performance. The simulations performed in this work using the ONE simulator have shown that information routing in social vehicle communities has better message delivery, less packet latency, and less data dropped percentage than that without social-aware communities. Thus, their associated carbon emissions have also been reduced because of all these benefits of communications efficiency.

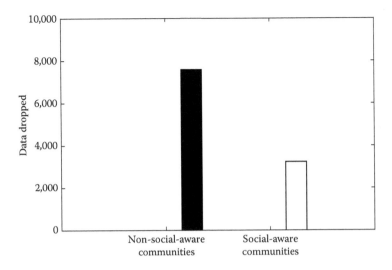

Figure 4.5 Comparison of data dropped in social-aware and non-social-aware communities.

Our future work will seek to further validate this social communities and transitivity-aware routing framework by using more complex and extensive network scenarios. It aims to achieve a high packet delivery ratio, lower overhead, and higher throughput for VANET dense-city scenarios.

References

1. U.S. Department of Transportation, Intelligent transportation systems (ITS), http://www.its.dot.gov/index.htm (last modified July 2015).
2. T.L. Willke, P. Tientrakool, N.F. Maxemchuk, A survey of inter-vehicle communication protocols and their applications, *IEEE Commun. Surv. Tutorials* 11 (2) (2009) 3–20.
3. Y. Toor, P. Muhlethaler, A. Laouiti, Vehicle ad hoc networks applications and related technical issues, *IEEE Commun. Surv. Tutorials* 10 (3) (2008) 77–88.
4. F. Terroso-Saenz, M. Valdes-Vela, C. Sotomayor-Martinez, R. Toledo-Moreo, A.F. Gomez-Skarmeta, A cooperative approach to traffic congestion detection with complex event processing and VANET, *IEEE Trans. Intell. Transp. Syst.* 13 (2) (2012).
5. F. Knorr, D. Baselt, M. Schreckenberg, M. Mauve, Reducing traffic jams via VANETs, *IEEE Trans. Intell. Transp. Syst.* 61 (8) (2012).
6. A. Busson, A. Lambert, D. Gruyer, D. Gingras, Analysis of intervehicle communication to reduce road crashes, *IEEE Trans. Intell. Transp. Syst.* 60 (9) (2012).
7. G. Karagiannis, O. Altintas, E. Ekici, G. Heijenk, B. Japan, K. Lizn, T. Weil, Vehicular networking: A survey and tutorial on requirements, architectures, challenges, standards, and solutions, *IEEE Commun. Surv. Tutorials* 13 (4) (2011) 584–616.
8. Y. Toor, P. Muhlethaler, A. Laouiti, Vehicle ad hoc networks applications and related technical issues, *IEEE Commun. Surv. Tutorials* 10 (3) (2008) 77–88.
9. N. Wisitpongphan, F. Bai, P. Mudalige, O.K. Tonguz, On the routing problem in disconnected vehicular ad-hoc networks, in *IEEE INFOCOM 2007, 26th IEEE International Conference on Computer Communications,* 2007, pp. 2291–2295.
10. V. Cerf, S. Burleigh, A. Hooke, L. Torgerson, R. Durst, K. Scott, K. Fall, H. Weiss, Delay-tolerant networking architecture, https://www.rfc-editor.org/rfc/rfc4838.txt (Internet RFC 4838, April 2007).

11. D. Câmara, N. Frangiadakis, F. Filali, C. Bonnet, in *Handbook of Research on Mobility and Computing*, M.M. Cruz-Cunha, F. Moreira (eds.), IGI Global, Hershey, PA, 2011, pp. 356–358.

12. L. Franck, F. Gil-Castineira, Using delay tolerant networks for Car2Car communications, in *IEEE International Symposium on Industrial Electronics, ISIE 2007*, vol. 2, no. 5, 2007, pp. 2573–2578.

13. V.N.G.J. Soares, J.J.P.C. Rodrigues, F. Farahmand, Performance assessment of a geographic routing protocol for vehicular delay-tolerant networks, in *IEEE Wireless Communications and Networking Conference*, 2012, pp. 2526–2531.

14. F. Li, Y. Wang, Routing in vehicular ad hoc networks: A survey, *IEEE Veh. Technol. Mag.* 2(2) (2007), 12–22.

15. J. Kurhinen, J. Janatuinen, Delay tolerant routing in sparse vehicular ad hoc networks, *Acta Electrotech. Inform.* 8 (3) (2008) 7–13.

16. S. Al-Sultan, M. Moath, Al-Doori, H. Al-Bayatti Ali, H. Zedan, A comprehensive survey of vehicular ad hoc network, *J. Netw. Comput. Appl.* 37 (1) (2014) 380–392.

17. I. Salhi, M. Cherif, S. Senouci, Data collection in vehicular networks, in *Autonomous and Spontaneous Networks Symposium*, 2008, pp. 20–21.

18. D.J. Watts, S.H. Strogatz, Collective dynamics of 'small-world' networks, *Nature*, 393 (6684) (1998) 440–442.

19. E. Schoch, F. Kargl, M. Weber, Communication patterns in VANETs, *IEEE Commun. Mag.* (2008) 119–125.

20. S. Panwai, H. Dia, Neural agent car-following models, *IEEE Trans. Intell. Transp. Syst.* 8 (1) (2007) 60–70.

21. M. Saito, J. Tsukamoto, T. Umedu, T. Higashino, Design and evaluation of inter-vehicle dissemination protocol for the propagation of preceding traffic information, *IEEE Trans. Intell. Transp. Syst.* 8 (3) (2007) 379–390.

22. S.-H. Lin, J.-Y. Hu, C.-F. Chou, I.-C. Chang, C.-C. Hung, A novel social cluster-based P2P framework for integrating VANETs with the Internet, in *IEEE Wireless Communications and Networking Conference*, 2009.

23. C. Shea, B. Hassanabadi, S. Valaee, Mobility-based clustering in VANETs using affinity propagation, in *IEEE Globecom*, 2009.

24. M. Drawil Nabil, O. Basir, Inter vehicle-communication-assisted localization, *IEEE Trans. Intell. Transp. Syst.* 11 (3) (2010) 678–691.

25. P. Wex, J. Breuer, A. Held, T. Leinmuller, L. Delgrossi, Trust issues for vehicular ad hoc networks, in *IEEE, VTC Spring 2008*, May 2008, pp. 2800–2804.

26. S. Milgram, The small world problem, *Psychol. Today* 1 (1) (1967) 60–67.

27. D. Saxbe, Six degrees of separation: Two new studies test 'six degrees of separation' hypothesis, *Psychol. Today* (2003).

28. W.X. Zhou, D. Sornette, R.A. Hill, R.I.M. Dunbar, Discrete hierarchical organization of social group sizes, *Proc. R. Soc. Lond. B Biol. Sci.* 272 (1561) (2005) 439–444.

29. S. Atev, G. Miller, N. Papanikolopoulos, Clustering of vehicle trajectories, *IEEE Trans. Intell. Transp. Syst.* 11 (3) (2010) 647–657.

30. E. Souza, I. Nikolaidis, P. Gburzynski, A new aggregate local mobility (ALM) clustering algorithm for VANETs, in *IEEE International Communications Conference*, 2010.

31. S.-S. Wang, Y.-S. Lin, Performance evaluation of passive clustering based techniques for inter-vehicle communications, in *19th Annual Wireless and Optical Communications Conference (WOCC)*, Shanghai, 2010, pp. 1–5.

32. Y. Xu, W. Wang, Topology stability analysis and its application in hierarchical mobile ad hoc networks, *IEEE Trans. Veh. Technol.* 58 (3) (2009) 1546–1560.

33. P. Fan, Improving broadcasting performance by clustering with stability for inter-vehicle communication, in *IEEE 65th Vehicular Technology Conference*, Dublin, 2007, pp. 2491–2495.

34. A. Daeinabi, A.G.P. Rahbar, A. Khademzadeh, VWCA, an efficient clustering algorithm for vehicular ad hoc networks, *J. Netw. Comput. Appl.* 34 (1) (2011) 207–222.

35. M. Bakhouyaa, J. Gabera, P. Lorenzb, An adaptive approach to information dissemination in vehicular ad hoc networks, *J. Netw. Comput. Appl.* 34 (6) (2011) 1971–1978.

36. Ford's vision, Innovation, http://corporate.ford.com/innovation/innovation-detail/pr-cars-talking-to-traffic-lights-and-34198 (last modified March 10, 2015).

37. F. Mezghani, R. Dhaou, M. Nogueira, A.-L. Beylot, Content dissemination in vehicular social networks: Taxonomy and user satisfaction, *IEEE Commun. Mag.* 52 (12) (2014) 34–40.

38. F. Xia, L. Liu, J. Li, A.M. Ahmed, L.T. Yang, J. Ma, BEEINFO: Interest-based forwarding using artificial bee colony for socially-aware networking, *IEEE Trans. Veh. Technol.* 64 (3) (2014) 1–11.

39. R. Fei, K. Yang, X. Cheng, A cooperative social and vehicular network and its dynamic bandwidth allocation algorithms, in *Proceedings of IEEE INFOCOM Workshops,* April 2011, pp. 63–67.

40. V. Loscri, A queue based dynamic approach for the coordinated distributed scheduler of the IEEE 802.16, in *Proceedings of IEEE ISCC,* 2008, pp. 423–428.

41. V. Loscri, G. Aloi, Transmission hold-off time mitigation for IEEE 802.16 mesh networks: A dynamic approach, in *Proceedings of IEEE WTS,* 2008, pp. 31–37.

42. A. Srivastava, Anuradha, D. Gupta, Social network analysis: Hardly easy, in *Proceedings of IEEE ICROIT,* February 2014, pp. 128–135.

43. G. Wang, W. Jiang, J. Wu, Z. Xiong, Fine-grained feature-based social influence evaluation in online social networks, *IEEE Trans. Parallel Distrib. Syst.* 25 (9) (2014) 2286–2296.

44. A. Papadimitriou, D. Katsaros, Y. Manolopoulos, Social network analysis and its applications in wireless sensor and vehicular networks, in *Next Generation Society: Technological and Legal Issues,* vol. 26, Lecture Notes Series of the Institute for Computer Sciences, Social-Informatics and Telecommunications Engineering, A. Sideridis, C. Patrikakis (eds.), Springer-Verlag, Berlin, 2010, pp. 411–420.

45. F. Cunha, A.C. Viana, R.A.F. Mini, A.A.F. Loureiro, Is it possible to find social properties in vehicular networks? in *Proceedings of IEEE 19th ISCC,* 2014, pp. 1–6.

46. V. Palma, A.M. Vegni, On the optimal design of a broadcast data dissemination system over VANET providing V2V and V2I communications—'The vision of Rome as a smart city,' *J. Telecommun. Inf. Technol.* 2013 (1) (2013) 41–48.

47. F. Cunha, A. Carneiro Vianna, R. Mini, A. Loureiro, Are vehicular networks small world? in *IEEE Conference of INFOCOM Workshops,* April 2014, pp. 195–196.

48. S. Panichpapiboon, W. Pattara-Atikom, A review of information dissemination protocols for vehicular ad hoc network, *IEEE Commun. Surv. Tutorials* 14 (3) (2012) 784–798.

49. A.M. Vegni, C. Campolo, A. Molinaro, T. Little, *Modeling of Intermittent Connectivity in Opportunistic Networks: The Case of Vehicular Ad Hoc Networks,* Springer-Verlag, New York, 2013.

50. R. Lu, Security and privacy preservation in vehicular social networks, PhD dissertation, University of Waterloo, Waterloo, Ontario, 2012.

51. F. Cunha, A. Carneiro Vianna, R. Mini, A. Loureiro, How effective is to look at a vehicular network under a social perception? in *Proceedings of the IEEE 9th International Conference of WiMob,* October 2013, pp. 154–159.

52. L. Maglaras, D. Katsaros, Social clustering of vehicles based on semi-Markov processes, *IEEE Trans. Veh. Technol.* 65 (1) (2016) 318–332.

53. A.M. Vegni, E. Natalizio, Forwarder smart selection protocol for limitation of broadcast storm problem, *J. Netw. Comput. Appl.* 47 (2015) 61–71.

54. A.M. Vegni, A. Stramacci, E. Natalizio, Opportunistic clusters selection in a reliable enhanced broadcast protocol for vehicular ad hoc networks, in *Proceedings of the 10th Annual Conference of WONS,* March 2013, pp. 95–97.

55. N. Lu, T. Luan, M. Wang, X. Shen, F. Bai, Bounds of asymptotic performance limits of social-proximity vehicular networks, *IEEE/ACM Trans. Netw.* 22 (3) (2014) 812–825.

56. N. Lu, T. Luan, M. Wang, X. Shen, F. Bai, Capacity and delay analysis for social proximity urban vehicular networks, in *Proceedings of IEEE INFOCOM,* March 2012, pp.1476–1484.

57. A. Vahdat, D. Becker, Epidemic Routing for partially connected ad-hoc networks, Technical report, Duke University, Durham, NC, 2000.

58. A. Lindgren, A. Doria, O. Schelen, Probabilistic routing in inter-intently connected networks, *ACM Sigmobile Mob. Comput. Commun. Rev.* 7 (3) (2003) 19–20.

59. O. Richters, T.P. Peixoto, Trust transitivity in social networks, *PloS One* 6 (2011) el8384.

60. H. Dissanayake, R. Koggalage, A cost-effective intelligent solution to reduce traffic congestion, in *Wireless and Optical Communications Networks (WOCN)*, 2010.

61. S. Lin, B. De Schutter, S.K. Zegeye, H. Hellendoorn, Y. Xi, Integrated urban traffic control for the reduction of travel delays and emissions, in *2010 13th International IEEE Conference on Intelligent Transportation Systems (ITSC)*, 2010, pp. 677–682.

62. K. Kraschl-Hirschmann, M. Zallinger, R. Luz, M. Fellendorf, S. Hausberger, A method for emission estimation for microscopic traffic flow simulation, in *2011 IEEE Forum of Integrated and Sustainable Transportation System (FISTS)*, June 29–July 1, 2011, pp. 300–305.

63. M. Barth, K. Boriboonsomsin, Real-world carbon dioxide impacts of traffic congestion, *Transp. Res. Rec.* 2058 (1) (2008) 163–171.

64. H. Li, Calculation of additional pollutant gas emissions and their social cost from transport congestion, presented at 2011 Second International Conference of Mechanic Automation and Control Engineering (MACE), July 15–17, 2011.

65. A. Keränen, J. Ott, T. Kârkkâinen, The ONE simulator for DTN protocol evaluation, in *Proceedings of the 2nd International Conference on Simulation Tools and Techniques, SIMUTools '09*, Rome, 2009.

66. M. Haklay, P. Weber, OpenStreetMap: User-generated maps, *IEEE Pervasive Comput.* 7 (4) (2008) 12–18.

67. OpenStreetMap to WKT conversion, http://www.tm.kit.edu/mayer/osm2wkt (last modified April 15, 2010).

68. L. Qianxi, L. Li, Low carbon transportation in Japan and its developmental analysis, in *6th Advanced Forum of Transportation of China*, 2010.

69. X. Chang, B.Y. Chen, Q. Li, X. Cui, L. Tang, C. Liu, Estimating real-time traffic carbon dioxide emissions based on intelligent transportation system technologies, *IEEE Trans. Intell. Transp. Syst.* 14 (1) (2013) 469–479.

70. M. Barth, F. An. T. Younglove, C. Levine, G. Scora, M. Ross, T. Wenzel. The development of a comprehensive modal emissions model. Final report submitted to the National Cooperative Highway Research Program, Nov. 1999, 255.

Chapter 5

Architectures for Social Vehicular Network Programmability

Flavio Esposito

Computer Science Department, Saint Louis University, St. Louis, Missouri

Contents

5.1 Introduction

For computer scientists and telecommunication engineers, vehicles are multi-radio-equipped moving processes, able to communicate with other internal or external (moving) processes. Vehicle innovation has been mostly dominated by mechanical engineers, while the telecommunication industry has not yet been able to capitalize on the vast potential of these ubiquitous mobile processes. By vehicles, we do not mean merely human-driven cars: wireless communications can occur among a wide range of mobile processes, such as subways, high-speed trains [30], boats, tractors, robots for precision agriculture, drones, and other unmanned aerial vehicles (UAVs). Most of these vehicles already embed software; some of them are even fully driven by software. For example, a few big airports already use unmanned trains to transport passengers across terminals. Software-driven vehicles, if well managed, have the potential to substantially reduce incident risks.

Moreover, large-capacity storage devices could be installed on any moving processing unit: trains, airplanes, trucks, or even commuting cars, enabling new ways of communication, new streams for advertising, and new safety or entertainment applications. In such an ecosystem, not only vehicles with their onboard sensors but also humans are data generators, publishers, and subscribers. Sharing and processing data while on the road (not necessarily while driving) can enable a number of helpful applications for drivers, as well as for public transportation passengers, road authorities, and transportation companies.

Not only do we know how to estimate vehicle mobility patterns [52,60,69], but also we know the schedules of trains, planes, buses, and subways. With the knowledge of where and when data could move, connected vehicles carrying (very large) storage units with a network interface may be the fastest data transfer and data sharing communication channel.

5.1.1 What Can We Share on Vehicular Networks?

Vehicles (not merely cars) may generate and publish useful information to subscribers [59]. In this chapter, we do not limit the notion of sharing to data, but we extend it to computations. Parked or moving vehicles may cooperate to mitigate the time to solution for expensive computations, for example, when the cloud computing infrastructure is unavailable, in some cases untrusted, or too costly.*

Cars, trucks, train, buses, tractors, or even drones may share a wide range of private or public collected or sensed information. Examples of such information include icy roads; heavy rain, snow, or fog; vehicle failures; geolocations; destinations or relative distance of a fleet of UAVs—in the case of coordination needs with other automated vehicles—traffic recommendations; voice notes; or even pictures of a scene, agriculture field area, or a target location (e.g., a fire to be extinguished, a road accident, roadwork, traffic congestions, or even road system failures).

---

* Examples of heterogeneous networks enabled by middleware that automatically or on demand switch to the lowest-cost network interface already exist [16–18].

5.1.2 Why VSN Programmability?

Historically, protocols have been designed in an ad hoc manner, and vehicular social networks (VSNs) are no exception. Programmability is key for evolution. To quickly adapt to new system, application, or user requirements, architectures and protocols for VSNs should be designed with adaptation capabilities. Defining an architecture means identifying the mechanisms in a given problem, and separating its functionalities. A mechanism is an invariance of a given problem, while a policy is a variant aspect of such a mechanism. For example, acknowledge is a mechanism, and when to acknowledge is a policy. Designing a programmable architecture for a VSN means identifying the adaptable mechanisms, for example, content discovery, forwarding, or routing, and separating such mechanisms from their policies. By allowing different policy instantiations (policy programmability), VSN protocol and distributed algorithms become more flexible; that is, they can easily adapt to the goals of different VSN service providers and applications. A VSN with reprogrammable (data transfer and data management) mechanisms also allows cheaper and easier monitoring, repair, and often even control of such networks, enabling faster, simpler, and more scalable data or processing sharing among vehicles.

5.1.3 Chapter Organization and Contributions

To (socially) capitalize on such data sharing or processing opportunities, and to design programmable VSN solutions, managers and application programmers need to cope with many architectural and environmental challenges. We describe some of these challenges in Section 5.2. In particular, we classify such challenges into (1) network management challenges (Section 5.2.1), (2) naming and addressing challenges (Section 5.2.2), and (3) application challenges (Section 5.2.3). The focus of this chapter is on programmability of the VSN management mechanisms. After discussing the VSN challenges, we hence describe three architectures whose concepts should be considered by designers of programmable VSN services and applications. In particular, in Section 5.3, we discuss a virtualization architecture that via software-defined networking (SDN) enables programmability of a few virtual network (VN) mechanisms hosting a VSN. In Section 5.5, we focus on delay or disruption-tolerant networking, an architecture that provides communication even when the end-to-end connections among vehicles are challenged by very large delays or by vehicles abandoning the network or unsubscribing to VSN updates. In the same section, we discuss the details of an approach to enable routing programmability within delay-tolerant network (DTN)-enabled vehicular ad hoc network (VANET) networks (Section 5.5.2). We then describe how future Internet architectures can help the design of VSN and we focus on a case study, the Recursive InterNetwork Architecture (RINA), in Section 5.6. Future Internet architectures are ambitious projects and an excellent educational exercise to understand the problems of the current Internet architecture, and to guide designers of novel architectures and applications for emerging technologies such as VSNs.

We then conclude the chapter by reporting what we believe are some of the most general and popular prototyping tools that are suitable for testing novel and existing VSN solutions.

5.2 Challenges in Providing a Social Vehicular Network Service

How can we support a *vehicle app store* where users download entertainment, safety, and vehicle maintenance applications? How can we leverage nearby vehicles for social distributed computation to achieve a common goal? How can we effectively create—allocate CPU and network resources—and maintain—dynamically adjust states as network conditions evolve—a scalable VSN?

Even when a VSN is established and maintained, promptly sharing large datasets or computation tasks brings a set of additional (Big Data) challenges. VSNs require dynamic access to large vehicle-generated datasets. How do we efficiently collect, store, and process such sensor-generated data with or without an infrastructure? Who should own and where should we process these massive datasets? Some applications could be locally managed by a single vehicle that subscribes to a service provider for updates; for other applications, instead we may need a set of federated infrastructure providers (InPs) cooperating (or competing) to provide a VSN service. In such large wide-area networks (WANs), how do we reach consensus on a state in a presence of (Byzantine) failures?

We broadly classify the challenges of providing a VSN service into three different categories: network management challenges, naming and addressing challenges, and application challenges. In the rest of this section, we detail some of these challenges, pointing out existing technologies and architectures that can be used to manage (and so monetize) this emerging type of networks. We also highlight a few limitations of such approaches, driving a need for novel solutions.

5.2.1 Network Management Challenges in VSN

By network management in this chapter, we refer to the mechanisms necessary to monitor, repair, and control a VSN. By control, we mean the possibility to edit any network state. Sometimes, in the network management literature authors separate network management and state control for some particular state, for example, flow control and congestion control.

A fundamental characteristic of a VSN is its dynamic nature. Not only the processes are mobile, but also users and vehicles may actively subscribe and unsubscribe to different (network) services or other users more frequently than on standard online social networks, given the relevance of locations and content to be shared.* VSNs need to create, destroy, and modify partially distributed databases that need to be kept consistent. Also, content needs to be generated, collected, and shared quickly, or it becomes irrelevant, misleading, or even dangerous in military applications. Selecting, routing, and scheduling content with different priorities poses additional challenges than those required to allow publishing on a standard social network where posts arrive in a *first come, first served,* fashion. Finally, in a VSN, the publisher location, the service, the application, and the urgency of the data to share play an important role in forwarding, authentication, enrollment, and state maintenance of the relevant content.

5.2.2 Mobility and Multihoming (Naming and Addressing) Challenges

Vehicles are mobile by definition, but they are also multihomed; that is, each node is connected to more than one network. Mobility can be viewed as a special case of multihoming—as vehicles move, they unsubscribe from one network and subscribe to another, which is akin to an interface becoming inactive and another active. The current Internet architecture has been facing significant challenges in effectively dealing with multihoming (and consequently mobility), which has led to the emergence of several custom point solutions. VSNs are no exception. There have been several attempts to fix this addressing problem, including the Location ID Separation Protocol (LISP)—tested at Cisco [29, 42]—and Mobile Internet Protocol (MIP) [8]. The basic idea behind LISP is to assign the multihomed node a provider-independent (location-independent) identifier (ID). A border router maps a destination ID to the node's location, which is the address of another border router that is known to have a path to the node. Routing is then done from the source's border router to the destination's border router. If the latter vehicles' location changes due to path failure

---

* By content, we mean any data that has a value for either a VSN user or a VSN service provider. It can be any sensed or human-generated data point.

or mobility, it becomes costly to propagate that change over the whole network (to all possible source border routers). MIP allows a mobile host to seamlessly move from its home domain to a foreign location without losing connectivity. This is done by having a foreign agent (router) update the location of the mobile node at its home agent (router). Since mobility is a special dynamic form of multihoming, MIP can also be used to handle a change in the active interface (due to failure or rerouting), leading to a multihomed node, where a home agent directs traffic to the currently active (operational or "better") interface. However, this location update can be costly since it needs to propagate from the foreign agent to the home agent. Note that both LISP and MIP (and the combination thereof) help reduce the size of the routing tables at the core of the Internet, since several IDs can map to one location and hence be represented by one routing entry. Further elaboration on the benefits of LISP can be found in [48].

Another recent alternative to solve the mobility and multihoming problem of any distributed overlay of processes is RINA. It uses the concept of distributed interprocess communication facility (DIF) to divide communication processes into manageable scopes across network subsystems, which results in a reduced routing table size per DIF (see Section 5.6 or [14, 23, 64] for details on the architecture). RINA processes a route hop-by-hop based on the destination's node addresses, not its interfaces. At each hop, the next-hop node address is mapped to the (currently operational) interface to that next-hop node. This late binding of a node's address to its interface (path) allows RINA to effectively deal with interface changes due to multihoming or mobility. The cost of such late binding is relatively small since its scope is local to the routing "hop" that traverses the underlying DIF. By recursing the DIF structure to make the DIF scopes small enough, the cost of such late bindings (location updates) can be made (in theory) arbitrarily small. More information on how RINA compares with different multihoming approaches is detailed in [34].

5.2.3 Application and Resource Management Challenges

Privacy is a challenge that VSN application programmers inherit from other social networks. We do not discuss privacy issues for VSN in this chapter, as others in this book do so. The interested reader can also find a good survey in Chapter 10 of [68] or in [5, 35, 54] and references therein. Similar considerations can be drawn for security; see, for example, [41, 49].

Data-intensive computing associated with VSN requires seamless processing of sensed data, for example, imagery or video, both at the network edge and at the core within the cloud. Such seamless processing is particularly critical in delay-sensitive applications, for example, online multiplayer games or when sharing data relative to a road incident. The vehicles generating (data, imagery, games, or video) content are rarely equipped with high-performance computation capabilities to run processing algorithms, for example, computer vision algorithms. To this end, a set of challenges arise in bootstrapping and keeping alive the virtual paths between the edge network and the cloud hosting the VSN computations. Some of these challenges include keeping a consistent state when forming the application overlay with consensus algorithms [39], or rapidly converging to a set of low-latency peers to establish an online gaming session among vehicle passengers [4].

Many other emerging applications demand complex computation and large storage. It becomes increasingly difficult for an individual vehicle to efficiently support these applications. A very promising solution is to share the computation and storage resources among all physically nearby vehicles. As such, a new paradigm has emerged called "vehicular cloud" [65]. A vehicular cloud consists of a group of vehicles whose corporate computing, sensing, communication, and physical resources can be coordinated and dynamically allocated to authorized users. Some literature refers to this concept (although not focused on vehicles) as fog computing. Here, the challenges may include collecting (sensing in a distributed fashion), computing (uploading to the cloud), and consuming (push the postprocessed data back after the cloud or the peer-to-peer processed data).

5.3 Virtualization Architectures for VSN

VSNs will use cloud, fog computing [58], SDN and software-defined infrastructure (SDI) tech-nologies. We all became familiar with the concept of cloud computing, where InPs share or lease a fraction of their physical network to a third-party enterprise or service provider with the aim of pro-viding a wide area (scalable) and elastic service.* The notion of cloud usually refers to virtualization of CPU resources, and often ignores virtualization of physical paths. Network virtualization tech-nology allows instead the creation and maintenance of instances of both nodes (physical machines) and links (loop-free physical paths) within data centers of at least one InP. OpenStack [46] is one popular example of an open-source project that allows virtualization of both nodes and links.†

Network virtualization has also enabled SDN. The general notion of SDN is to centralize the management of network state updates by offloading the logic of a mechanism (often merely the forwarding mechanism) to an external off-device controller. All SDN models have some version of network service controller, as well as a southbound and a northbound application programming interface (API) (Figure 5.1). A northbound API allows programmability of social vehicular net-work applications, defining high-level rules such as those defined by business logic. A southbound API instead allows programmability of the underlying (historically hardware) network compo-nents, such as the router, switches, or firewalls. More recently, with the advent of network function virtualization (NFV), researchers have proposed to replace some of the underlying hardware func-tionalities, such as firewalls or load balancers, with their equivalent software-defined version for better programmability.‡ OpenFlow [40] was the original southbound API and remains one of the most common protocols to control the forwarding mechanism. These two APIs help network administrators to programmatically shape traffic and deploy VSN services across a wide range of InP.

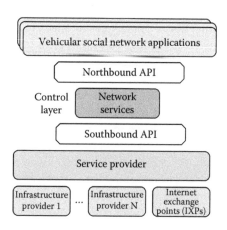

Figure 5.1 SDN hourglass model: The control of a few essential network services, combined with the northbound and southbound APIs, is enough for a service provider to manage the state of multiple VSN applications across a range of multiple InPs and IXPs.

---

* Elasticity is the ability to rapidly scale up and scale down network or computing resources.
† In the literature, the notion of software-defined environment (SDE) is also present, where not only the network but also the database and the applications are virtualized.
‡ A recent and interesting line of research focuses on NFV via Field Programmable Gate Array (FPGA), keeping the network functionalities in the hardware but enabling programmability; see, for example, [36, 66].

5.4 Software-Defined Networking and Flow Controllers

Although controllers were designed to be the logically centralized elements to dictate policies for the forwarding mechanism only, the notion of controller is now extended as the *brain of the network*; that is, it may implement more general software able to instantiate policies for other VSN mechanisms as well. Some of these programmable network management mechanisms that would make VSNs easier to manage include routing [10, 20, 61], scheduling [55], and VN embedding [12,19,21,22]. By having a (logically) centralized view of the entire network, controllers enable network administrators to easily express high-level VSN rules, for example, route all images to node X for a report on traffic conditions into low-level policies, that is, the underlying network system that handles network traffic, for example, a set of Linux *iptables* rules [33].

Although software-defined networks have gained popularity during the past few years, many of the ideas underlying SDN are at least two decades old. The idea of separating control and data planes to simplify network management and the deployment of new services is merely revisited from early telephony networks. Controllers such as OpenFlow [40] rendered controllability easier where it was impossible on closed networks designed for a narrower range of telephony services. Also, the vision of programmable networks had been floated before—even if, with an emphasis on programmable data planes, programmable networks were already present in the *active network* research community. Finally, SDN also relates to previous work on separating the control and data planes in computer networks. A more detailed history on the SDN evolution is illustrated in [26].

To conclude, a network virtualization layer orchestrated by an SDN controller would allow programmability of several network management mechanisms with the aim of fostering cooperation between the (wireless) network, the wired network of infrastructure and service providers, and its upper social application layers. By social application layer, we mean that VSN application programmers may also leverage SDN for computation sharing or offloading. The virtualization paradigm that can be found in the mobile cloud computing literature assumes only virtual machine usage [15,27]. In VSN applications, instead we need to deal with resource allocation and management of the VNs between the fog (nearby virtualized nodes), the cloud sites (remote virtualized nodes), and the processes on the vehicle itself (local processes).

In the rest of this section, we give an example of a programmable VN management mechanism that VSN managers may leverage to tune their high-level policies, adapting to the goals of providers and VSN applications. In particular, we discuss in Section 5.4.1 the *VN embedding*, the NP-hard graph matching problem of creating a virtual (social) network on top of a constrained physical network.

5.4.1 Vehicular Social Network Embedding Programmability

We have learned so far that network virtualization and SDN technologies have enabled multiple virtual instances of a VSN to coexist on a common (wireless) physical network infrastructure. Each VN is customizable in support of a wide range of social applications. If the VSN is designed to enable policy programmability, then managing it becomes easier, as application programmers can use a single (northbound or southbound) API to adapt to different applications and scenarios, without having to implement from scratch a new protocol for each new application. One of the fundamental management protocols, not yet standardized, that VSN providers need to run to support such vehicular social services is the VN embedding protocol (Figure 5.2).

Running such a protocol requires solving the NP-hard problem of matching constrained VNs on the physical network (overlay), owned by a single provider or by multiple federated providers. The VN embedding problem consists of three interacting (and customizable, or policy-programmable) mechanisms: (1) resource discovery, where the space of available, potentially

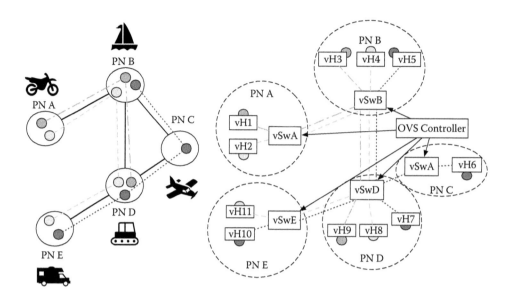

Figure 5.2 Many vehicle processes may be subscribed to multiple social networks. A VSN is logically a VN. A centralized open virtual switch (OVS) controller may steer traffic from each software-defined node to optimize content sharing and push relevant information in a timely fashion.

hosting resources (within vehicles) is sampled or exhaustively searched; (2) VN mapping, where a subset of such available resources is chosen as a candidate to potentially host the requested (VSN) application; and (3) allocation, where each virtual node is bound to at least a physical process in the underlying physical network, and each virtual link to at least one loop-free physical path.

Many centralized VN embedding solutions focus on specific policies under various settings. Policies (i.e., high-level goals) parametrize any of the three embedding mechanisms. For example, some centralized heuristics devised for small enterprise physical networks embed virtual nodes and virtual links separately to adapt the load on the physical network resources with minimal virtual machine or path migrations [67]. Other solutions have shown how the physical network utilization may increase by simultaneously embedding virtual nodes and links [12]. Distributed solutions for embedding wider area VNs also exist [13,19,28,70]. Some of them outsource the embedding to a centralized service provider that coordinates the process by either splitting the VN request and sending it to a subset of InPs [28], or collecting resource availability from InPs and later offering an embedding [70]. Outsourcing the embedding has the advantage of relieving InPs from the entire VSN management complexity, but a single centralized authority is nongeographically distributed, and could be untrusted, or a single point of failure. On the other hand, centralized solutions are more suitable for smaller-scale social network providers like Bonfyr [6], a small event-based social network, where controllability is more important than scalability. To our knowledge [21], none of the existing embedding solutions are designed to easily adjust to the goal of a socially aware network application with the constraints of a VSN service or infrastructure provider. Two examples of policy-based VN embedding design can be found in [13] and [19].

In [13], the authors propose a VN embedding protocol that allows InPs to pick their own utility and declare their own cost when attempting an embedding; then the service provider picks the lowest-cost embedding solution. In [19], the authors propose a consensus-based auction with guarantees on the VN embedding time and performance, with respect to a Pareto optimal embedding. In this latter case, the VN embedding policies, which can be reprogrammable, are not limited to the provider utility function, but include several node and link policies as well, for example, how many virtual nodes each physical node is allowed to bid on per auction round, or the value of k in the k shortest-path-link embedding algorithm.

5.5 Delay- or Disruption-Tolerant Architectures for VSN Programmability

VANETs in general and VSNs in particular are characterized by intermittent connectivity as vehicular traffic conditions may drastically change with time and space. Any application over a vehicular network, especially VSNs, can and should be considered a DTN. The notion of a DTN was exposed for the first time at Intel Labs, Berkeley (formerly called Intel Research Berkeley and Future Technologies Research Berkeley) in 2003 [25], with an Internet Engineering Task Force (IETF) draft that later became a Request for Comments (RFC) [9].*

The work on DTN originated from a research project in the Jet Propulsion Lab (JPL) at NASA, called InterPlanetary Networking (IPN) [7, 31]. The main scope was to create a communication protocol between the earth and other vehicles in space. The goal was to extend the current Internet architecture to environments with high delay and communication interruptions (typical in satellite scenarios).

After the JPL, other research groups, including VANET researchers, began to be interested in DTNs, as the same concepts can be applied to any network of sensors in challenged environments. Most of these researchers formed the DTN Research Group (DTNRG). The IPN (NASA) group and the U.S. Defense Advanced Research Projects Agency (DARPA) also carry on research on DTNs, but DTNRG is the only open research group. Originally, the term *disruption* to the acronym DTN was introduced by DARPA for military applications. Vehicles abandoning the range of operation of their (wireless) social networks can also be modeled as disruptions of the VSN.

Standard network protocols developed for the Internet may not work in challenged networks like VSN: both transport and routing protocols may fail. In some cases, even if they may experience poor performance, in the presence of intermittent connectivity or long propagation delays many Internet protocols still work. However, applications requiring reliability will not function properly. Many applications designed to operate on the Internet rely on transport protocols to guarantee a 100% packet delivery rate.†

The vast majority of Internet traffic over an IP layer uses Transmission Control Protocol (TCP) and User Datagram Protocol (UDP). UDP is unreliable by design. Typically, TCP has a negotiation phase to regulate the data flow between the source and destination, and to estimate the connection latency. If the actual latency is grater than what is estimated, transmissions are likely to fail. Techniques such as *selective repeat* in TCP improve the throughput, but not when the end-to-end delay is drastically high, due to latency, network congestion, or the nature of the VSN application.

Routing table formation algorithms also rely on updated topologies, and if such information arrives with high delays, packets are likely to be routed over obsolete or inexistent paths. Moreover, packet delivery is likely to fail when delays cause the routing tables to partition the underlying physical network. For those reasons, a DTN may be necessary when designing and implementing VSN applications.

---

* K. Fall received in 2013 the Test of Time Paper Award [3] for his seminal publication on DTN [25].
† These guarantees are regulated by service level agreements, and they are not theoretical. Practically, to increase reliability, system engineers replicate data over different channels and remove duplicates at the destination, de facto reducing substantially losses at the expense of overprovisioning.

5.5.1 DTN Bundle Protocol

The key DTN idea, formulated in the IPN project by NASA, was an Internet-independent middleware between the transport layer and the application layer. Using the capabilities of the underlying protocols, the so-called *bundle protocol* allows a reliable end-to-end transmission even in networks challenged by high delay and disruptions, like some VSNs. With respect to the International Organization for Standardization Open Systems Interconnection (ISO-OSI) layering architecture, the bundle layer, that is, the bundle protocol and the convergence layer, combines functionalities of the presentation and the session layers.

In a VSN, application programmers could leverage the bundle protocol store-and-forward feature to have vehicles publish updates on traffic, accidents, blocked roads, or even severe weather conditions, such as sudden fog or icy roads, to vehicles traveling in the opposite direction, quickly reaching other vehicles behind, suggesting a detour or a speed reduction, way faster than a classical frequency modulation radio traffic update does today.

The bundle layer forms a binding store-and-forward overlay that allows communication between two or more regional (vehicular social) networks with limited scope, that is, range of operation (Figure 5.3). The bundle layer is made by two architectural components: The bundle protocol and the convergence layer adapter (CLA), often called merely convergence layer for abuse of language.

The difference in routing algorithms is another obstacle that hinders communications and data sharing among different VSN nodes. A VSN with enabled DTN capabilities solves such routing integration issues by having two identifiers in border gateways: a *region ID*, which identifies the regional network, and an *entity ID*, which identifies each node within a region. Only gateways located in between regions may identify the destination region, while the entity ID is used to locate the destination once the regional network has been located.

Together with the store-and-forward property, this routing hierarchy is ideal for social vehicular networks, as it ensures end-to-end connectivity even though, before starting the communication, neither connectivity nor a complete addressing resolution is established. While the former hierarchical routing property is also common in the Internet routing with the Border Gateway Protocol (BGP), that is, with the external BGP (eBGP) and internal BGP (iBGP) separation, the latter property, called *late binding*, is not currently used in the Internet, where a network address translator (NAT) and a Domain Name Server (DNS) need to be adopted to route packets to regional (private) networks.

Late binding has also been adopted more recently in solutions attempting to fix the mobility and multihoming problems of the current Internet architecture (see Section 5.2.2).

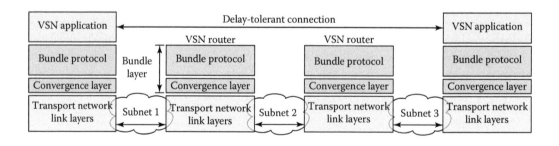

Figure 5.3 The bundle layer forms a binding store-and-forward overlay that allows communication within two or more regional VSNs with limited scope—range of operation.

In the rest of this section, we show an example of how the convergence layer can be augmented to support programmability of a social vehicular network by reporting a case study on Predicate Routing for DTN Architectures (*PreDA*) [20]. Its design goal is to unify mobile ad hoc network (MANET) architectures such as a VANETs with the DTN architecture by augmenting the bundle layer and by providing an API for routing programmability.

5.5.2 PreDA: An Example of Delay-Tolerant Network Programmability

Before SDN and OpenFlow [40] became popular in the network research community, few attempts at making network management mechanisms more programmable by injecting high-level rules from an orchestrator node were made. One of these examples is predicate routing [51]. Quoting Roscoe et al. [51], predicate routing defines *"the state of the network declaratively as a set of boolean expressions associated with links which assert which kind of packet can appear where."*

For a VSN user, a predicate is a high-level constraint on the routing of the content to be shared, injected by any vehicle supporting DTN capabilities; for example, direct all images captured by the camera on vehicle S to another (vehicle or infrastructure) intermediate node I for Big Data preprocessing, authorization, or filtering, before sending them to another VSN user destination D.

To support predicate routing in a MANET in general, and a VANET in particular, with or without social network capabilities, the system needs to map declarative user policies to network-level forwarding rules. In [20], predicates get propagated and installed as MANET-level forwarding rules. Any DTN-enabled VSN node can inject, from the application level, a routing rule (similarly to a post on typical social media) that gets seamlessly translated into a VANET-level rule, enabling new MANET or VANET routing instructions, that is, forwarding rules.

We next show two examples (Tables 5.1 and 5.2) of routing predicates.

Consider the injection of two predicates as in Table 5.1. The objective of these two rules is to redirect traffic destined to a vehicle D to an intermediate DTN node I for authentication or preprocessing. The first rule directs all data destined to vehicle D but not yet prescreened at the infrastructure or intermediate node I to node I.

Table 5.1   Direct all D Traffic to an Intermediate DTN Node I for Content Authentication

Predicate	Action
src = $\neg I \wedge$ dest = D	To I
src = $I \wedge$ dest= D	To D

Note: The second predicate allows packets to leave again after they have been rerouted to the intermediate node.

Table 5.2   Drop Traffic Not Originated by a White List W of VSN Nodes and Directed to a Private Node

Predicate	Action
src = $\neg W \wedge$ dest= D	Drop

Note that if the destination vehicle D is in the path to reach the intermediate node I, then vehicle D forwards the MANET or VANET packets matching this rule without reassembling the associated data message for prescreening.

The second predicate ensures that prescreened data coming from the intermediate DTN node I reach the original destination D.

Note that if nodes (S, I, and D) are DTN nodes, then these predicates override the normal DTN routing process. In particular, under normal DTN routing, node I, not recognizing itself as the destination, would have directed received bundles to D without preprocessing. Thus, the PreDA architecture supports predicate routing at the DTN level as well.

Consider now the injection of the predicate in Table 5.2. In this second use case, a predicate drops all the unsafe traffic coming from a *black list* of IP addresses directed to a private node D. In both cases, nodes not yet aware of the injected predicate follow the normal routing behavior.

5.5.3 PreDA Architecture

Figure 5.4 shows the DTN-MANET stack—the modified and added components to mimic a VSN support are marked by "stars."

The block named API of the DTN reference implementation is extended to allow applications to inject high-level routing requirements or constraints. The authors in [20] refer to this modification as *predicate routing API* (PR-API).

The *predicate routing support code* (PRSC) component, implemented in the routing protocol user space, mainly implements two functionalities: (1) it uses the *iptables* Linux facility [33] to install predicate MANET routing rules, so that VANET or MANET packets carrying VSN content to share—IP packets or DTN bundles—are routed based on DTN-level routing constraints, and (2) it creates and manages new routing extensions to discover other DTN nodes and propagate VANET or MANET routing predicates.

The *convergence layer* interfaces DTN and VANET or MANET by maintaining the mapping between DTN node names and IP addresses. The mappings are used to translate routing predicates on DTN node names to routing predicates on corresponding IP node addresses.

5.5.3.1 Predicate Routing Support Code

Dissecting the main architecture components modified and integrated together, in a bottom-up approach, we start by describing in detail the PRSC.

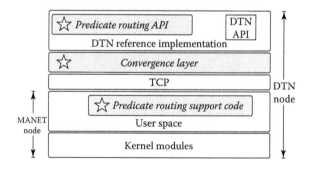

Figure 5.4 Architecture of a DTN node running over a VANET or MANET substrate. The VSN node is predicate routing enabled.

The PreDA architecture allows every node to behave as a pure VANET or MANET node or, when needed, to use the DTN properties. This flexibility is achieved by configuring on demand the properties that the user or the vehicular social network application wants to embed in such nodes. Listing 5.1 lists the options that the routing protocol daemon of PreDA supports for the case of the Ad Hoc On-Demand Distance Vector (AODV) protocol, chosen as a use case for the PreDA prototype implementation.

All the VSN nodes have to have a way to advertise two types of information: (1) DTN discovery and (2) predicate routing. In the particular case of AODV, this information is piggybacked as additional extensions to the HELLO message. The daemon needs to be launched with options –e and –p to allow other nodes to discover it. If option –a is active (active predicate routing), then users can also set the HELLO message frequency by selecting to send the predicates around the neighborhood every *N* seconds with option –b.

Listing 5.1   Routing Protocol Daemon Options

```
##    DTN    Discovery   ##
-e,           —use—dtn—eid           EID
-p,           —use—dtn—port          PORT
##    Predicate    Routing   ##
-a,           —predicate—enabled
-b,           —discovery—HELLO       N
```

Information for DTN discovery consists of, among other details that we omit in this discussion, (1) the DTN *endpoint identifiers* (EIDs) or addresses, specified in a Uniform Resource Identifier (URI)-style format, and (2) the port that the EID will use for its communication. This information gets propagated over the network to other nodes, if the node runs the DTN reference implementation.

On the other hand, information for predicate routing needs to be supported and advertised in every vehicle. In the AODV example shown in Listing 5.1, routing predicates are attached into new HELLO message extensions (with option –a), and they get disseminated every *N* seconds (option –b).

5.5.3.2 Convergence Layer

The convergence layer manages connections and interprocess communications (IPC) between the *DTN bundle layer* and the underlying transport protocol required by the VSN application. The PreDA convergence layer accomplishes the following two tasks: (1) manages and to maintains an up-to-date DTN-MANET/VANET address mapping (e.g., Table 5.3), and (2) converts declarative routing instructions coming from the DTN bundle layer into imperative *iptables* rules [33] (details

Table 5.3   DTN-VANET Address Mapping

DTN EID	IP Address	Port
dtn://vehicle1.dtn	10.0.0.1	5000
dtn://vehicle2.dtn	10.0.0.2	5000
dtn://vehicle7.dtn	10.0.0.7	5000
dtn://vehicle10.dtn	10.0.0.10	5000

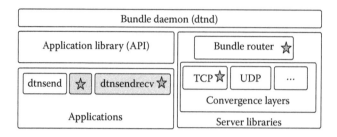

Figure 5.5 Architecture of a DTN reference implementation. The modules enabling predicate routing are marked with a star. Darker blocks represent applications enabled by PreDA.

in [20]). Note that if a predicate routing rule is created to redirect some traffic, then associated control messages, for example, keep-alive messages, would also be redirected.

5.5.3.3 Predicate Routing API

In Figure 5.5, we zoom in on the DTN reference implementation and PR-API (Figure 5.4). The modified and added components are marked by "stars." Blocks on the left half of the figure (*dtnse-crecv*, etc.) represent places where PreDA users would add VSN applications, such as the content sharing. Other blocks constitute our PreDA modification to the DTN reference implementation that users do not need to modify while embedding their VSN applications. A fundamental DTN characteristic is that when a (VSN) node receives a bundle, it is able to store it and wait until a valid path to the destination is found or restored. When a bundle arrives at a DTN node *I*, which is not the final destination, the bundle enters the so-called *pending state*. The bundle router, handling the bundles in the pending state, is responsible for the configuration of DTN routing. This module had to be modified to support PreDA applications that require predicate routing at the DTN level as well, where a data bundle received by an intermediate DTN node can be processed *before* being forwarded to its final DTN destination.

5.6 Future Internet Architectures

Around July 2009, the U.S. National Science Foundation (NSF) launched FIND [57] and a few funding initiatives to foster research in future Internet architecture design. As a result, many universities grouped their efforts to propose new solutions that would overcome the impasse reached by the current Internet architecture. Some examples are RINA [50], Named Data Networking [44], MobilityFirst [43], NEBULA [45], ChoiceNet [11], and XIA [24]. A VSN application can probably run on any of these existing architectures. As the author of this chapter directly contributed to the design and prototyping efforts of RINA, the focus in this section is on RINA [50], and on how it could be used to design and implement applications to manage VSNs by management policy programmability.

5.6.1 Recursive InterNetwork Architecture

One of many fundamental problems with the current Internet architecture is the lack of *scoping* of control and management functions, which makes it challenging to deliver a communication service with required characteristics when the range of operation (*e.g.*, capacity, delay, and loss) is so wide.

This problem is exacerbated in VSNs, where the underlying network is highly dynamic, datasets to be processed can be very large, and the number of subscribers and publishers may change rapidly. Also, a VSN may be a collection of private networks, that is, managed by a logically centralized entity.

RINA [14,23,34,62–64] is an architecture based on the fundamental principle that *networking is IPC and IPC only*. An IPC service is the RINA building block that provides communication service. RINA recurses the IPC service over different scopes, that is, ranges of operation. Specifically, a collection of distributed IPC processes with shared states is called a *DIF*. A DIF *layer* provides data transport services over a certain scope. In RINA, the concept of DIF is very powerful, since every private, virtual, or social network can be subsumed by a DIF. By *DIF layer instantiation*, we mean that a layer is built by customizing management and data transfer control policies to deliver predictable services to (VSN) applications. Stacking DIF layers on top of each other allows networks to be built from smaller and more manageable layers of limited scope. The DIF layer instantiation is *dynamic*, as layers can be discovered and created on the fly, and their mechanisms are able to respond to and support policy adaptation as network states (*e.g.*, quality of underlying services) change.

RINA enables private (*e.g.*, virtual or content delivery) networks to be dynamically instantiated, by customizing network management policies, for example, authentication, routing, addressing, and resource allocation, into a single layer, without the shortcomings of the TCP/IP architecture. For example, since the DIF layer identifies applications (services) by location-independent names, they can move, migrate, or be multihomed; since RINA supports explicit (authenticated) enrollment into the DIF layer, communication through the layer is secure; and since DIF services are recursively built, they can be better coordinated and managed. RINA does not have layers as in the ISO-OSI architecture; each DIF may implement the functionalities present in all layers of the ISO-OSI architecture. In [23], the authors demonstrated how a DIF can be built to provide a communication service for any application running on a private network. So in that case, a DIF only contains the mechanisms of a virtual private network.

One of the RINA architectural components is the so-called resource information base (RIB). Just as a forward information base (FIB) that stores all of the forwarding states, or a standard RIB that stores routing states in every router, the RIB keeps track of all states necessary for enabling communication in a VSN DIF. The RIB is a partially replicated distributed database updated with a publish/subscribe system that maintains it. When a new node (vehicle) joins a DIF, it subscribes to the (management) state information in use, as well as to its policies. At the base of any VSN, there is a pub/subsystem. Any VSN can be represented by a DIF or implemented within a DIF, since any process within any vehicle uses IPC to provide a communication social service. The advantage of using a RINA DIF to implement a VSN is maximum when different subnetworks operate over different network and application policies, for example.

5.7 Testing and Prototyping VSN Applications

In the past sections of this chapter, we have seen examples of architectures that can be used to design VSN applications and protocols. The networking and distributed system community is becoming skeptical of the significance of research results based solely on simulations. This fostered extensive research oriented toward networked emulation and prototype tools, in order to help perform realistic but reproducible (wide area) network experiments. More importantly, to utilize and demonstrate a new technology, implementation is key. In this section, we describe a few popular and novel but promising tools for evaluating and prototyping VSN protocols and applications. When implementing or simulating a VSN, researchers could choose among several other tools.

We report what we believe are some of the most general tools that are suitable also for testing VSN solutions. These tools can be classified into simulators, emulators, simulators with emulated interfaces, and experimental protocols with interfaces for rapid prototyping.

5.7.1 Mininet and MiniNExT

Mininet is an open-source network emulator [38], developed to foster rapid prototyping in the context of SDN. Recently, Mininet was extended by researchers at the University of Southern California to MiniNExT [53], where larger and more complex networks can be modeled, including IXPs.* In a scenario where a VSN is provided by sharing the physical infrastructure of multiple providers, MiniNExT could be used to prototype (VN) distributed data sharing protocols specifically oriented toward vehicular applications.

With the help of the Mininet or MiniNExT API, researchers can test their own protocols or architectures on emulated virtual graphs. It is possible to create realistic VNs with hosts, switches, routers, and links, reserving virtual node and link capacity. All VSN networks would be contained in a single hosting machine, and each virtual node (vehicle) may run real (Linux) applications such as ping, traceroute, or iperf [32]. Moreover, Mininet and MiniNExT enable experimenters to customize their network graph, including CPU and bandwidth constraints on the virtual nodes and links, and allow multiple developers to work on the same graph at the same time.

5.7.2 Click

Click is an open-source (both user and kernel space) tool that can be leveraged to build software for programmable and configurable routers [37]. Click configurations are modular and easy to extend, thus providing a flexible platform for VSN protocol development, testing, and deployment. A Click router is composed of packet processing modules called *elements*. Such elements enable a fine-grained control over the forwarding path. A Click router configuration is a directed graph that uses such elements as vertices. Each element implements a simple router function. A possible path of packets is represented by connections between pairs of elements, and packets flowing along the edges of such a graph. The Click implementation supports different packet scheduling and dropping policies, several queuing policies, and other packet processing applications.

5.7.3 OpenFlow

OpenFlow is an open standard that enables researchers to run experimental protocols [40]. The term *OpenFlow* refers to the standard, the protocol, and the controller that injects different forwarding policies. The OpenFlow protocol relies on Ethernet switches, and provides standardized interfaces to add and remove flow entries on a switch's internal flow table. With the OpenFlow standard, researchers can run experiments without knowing the internal details of a vendor's network devices. An OpenFlow-enabled switch separates packet forwarding (data plane) and routing (control plane) decisions. The data plane still resides on the switch, but the control plane is moved to a separate (OpenFlow) controller. The OpenFlow switch and controller communicate in user space through the OpenFlow protocol.

---

* The primary purpose of an IXP is to allow networks to interconnect directly, via the exchange, rather than through one or more third-party networks. The advantages of the direct interconnection are numerous, but the primary reasons are cost, latency, and bandwidth. IXPs are fundamentally reshaping Internet topologies, as well as its economic and peering relationship.

5.7.4 Network Simulator

Network simulator 3 (ns-3) advertises itself as an open-source discrete event network simulator, which allows studying Internet protocols and large-scale systems in a controlled environment for research and educational purposes [2]. Unfortunately, ns-3 is not backwards compatible with its widely diffused predecessor, ns-2 [1]. Ns-3 design focus has been on improving the core architecture of ns-2, enabling integration with other software for both research and educational applications. Ns-3 maintains an extensible software core, and has a flexible tracing system. Its implementation supports interfaces such as sockets API and Linux device driver. It also supports better software integration than ns-2, allowing integration with other open-source tools and models. For example, ns-3 has an interface to the Click modular router [37], and an interface to OpenFlow [40] to allow the simulation of virtual routers and switches within a VSN. Ns-3 supports two modes of integration with real systems: (1) virtual machines running on top of ns-3 communication channels and (2) an ns-3 stack over real devices.

5.7.5 Optimized Network Engineering Tool

Optimized Network Engineering Tool (OPNET) is a commercial network simulator that provides a VN environment able to model a complete network, including hosts, routers, switches, protocols, and applications [47]. OPNET is a commercial software, but is free to qualifying universities worldwide for academic research and teaching. OPNET provides a hierarchical graphical user interface (GUI)-based editor (network, node, and process), and supports discrete event simulation and flow analysis: a discrete event simulation allows packets and protocol message simulations, and the flow analysis enables modeling steady-state network behavior. OPNET also provides interfaces for live integration with the network system and other simulators.

5.7.6 ProtoRINA

In RINA [64], VSN network or applications innovation can be created by leveraging the RINA API, and such solutions can be tested using the existing user space [64] or kernel space [56] prototype. To test existing supported policies in their own customized environment, researchers can add novel policies or leverage the RINA communication service to write external VSN applications. According to the authors, the unique features of ProtoRINA [64] are

- **Customizable scope:** A DIF can be quickly and easily set up and customized by leveraging the provided mechanisms of the framework. Configurable policies include instantiation of several management mechanisms, for example, routing or enrollment. None of the existing simulation and emulation tools enable researchers to test different routing policies in different private networks (DIFs), each within a given limited scope.

- **Policy-based networking:** RINA supports SDN. Users can set up a private (social or virtual) network, merely as a high-level DIF on top of existing physical networks (lower-level DIFs), and instantiate a variety of mechanisms, rather than just forwarding, as in OpenFlow.

- **Recursion:** Another feature that makes RINA unique in the space of testing tools is recursion. Both data and control traffic are aggregated and transferred to lower IPC layers, and each VSN DIF may regulate and allocate resources to user traffic originating from higher-level DIFs or applications.

Given the above features, and the fact that RINA already has implemented a pub/sub system, ProtoRINA appears to be an interesting solution for VSN programmability.

5.8 Conclusion

In this chapter, we presented some challenges that network managers and application programmers may encounter when building and maintaining a VSN service. In particular, we classified such challenges as (1) network management challenges, (2) naming and addressing challenges, and (3) VSN application challenges. We then introduced three architectures that can be leveraged to support VSN programmability. First, we discussed how network virtualization and SDN can help reprogram some of the VSN management mechanisms, such as by solving a VN embedding problem during the formation of a VSN, and after the VSN is created, by steering (changing forwarding policies) collected and posted traffic with a centralized (OpenFlow) controller. Second, we pointed out how VSN applications may need to be tolerant to high delays and severe disruptions; that is, frequent change of subscribers, and we described the DTN architecture, as well as a concrete example of routing programmability within delay-tolerant MANET or VANETs. Finally, we described a more revolutionary approach to design VSN applications, that is, using RINA, a new Internet architecture based on the principle that networking is IPC.

Finally, we closed the chapter with a section on popular and innovative tools that can be used to prototype protocols and applications for VSNs.

References

1. The network simulator—ns-2. http://www.isi.edu/nsnam/ns/.
2. The network simulator—ns-3. http://www.nsnam.org/.
3. ACM SIGCOMM Test of Time Paper Awards. 2015. http://www.sigcomm.org/awards/test-of-time-paper-award
4. Sharad Agarwal and Jacob R. Lorch. Matchmaking for online games and other latency-sensitive p2p systems. In *Proceedings of the ACM SIGCOMM 2009 Conference on Data Communication*, SIGCOMM '09, pp. 315–326. New York: ACM, 2009.
5. Tracy Ann Kosa, Stephen Marsh, and Khalil El Khatib. Privacy representation in VANET. In *Proceedings of the Third ACM International Symposium on Design and Analysis of Intelligent Vehicular Networks and Applications*, DIVANet '13, pp. 39–44. New York: ACM, 2013.
6. Bonfyr Inc. https://bonfyreapp.com/.
7. Scott Burleigh, Adrian Hooke, Leigh Torgerson, Kevin Fall, Vint Cerf, Bob Durst, Keith Scott, and Howard Weiss. Delay-tolerant networking: An approach to interplanetary Internet. *IEEE Commun. Mag.*, 41(6):128–136, 2003.
8. Ed Charles Perkins. IP mobility support for IPv4. RFC 3344. August 2002.
9. Vint Cerf, Scott Burleigh, Adrian Hooke, Leigh Torgenson, Robert Durst, Keith Scott, Kevin Fall, and Howard Weiss. *Delay tolerant networking architecture*. RFC 4838. IETF DNT Research Group, April 2007.
10. Chao-Chih Chen, Lihua Yuan, Albert Greenberg, Chen-Nee Chuah, and Prasant Mohapatra. Routing-as-a-service (RAAS): A framework for tenant-directed route control in data center. *IEEE/ACM Trans. Netw.*, 22(5):1401–1414, 2014.
11. Choicenet: Network Innovation through Choice. 2010. https://code.renci.org/gf/project/choicenet/
12. Mosharaf Chowdhury, Muntasir Raihan Rahman, and Raouf Boutaba. ViNEYard: Virtual network embedding algorithms with coordinated node and link mapping. *IEEE/ACM Trans. Netw.*, 20(1):206–219, 2012.

13. Mosharaf Chowdhury, Fady Samuel, and Raouf Boutaba. PolyViNE: Policy-based virtual network embedding across multiple domains. In *Proceedings of ACM SIGCOMM Workshop on Virtualized Infrastructure Systems and Architecture* VISA '10, pp. 49–56. New York: ACM, 2010.

14. John Day, Ibrahim Matta, and Karim Mattar. "Networking is IPC": A guiding principle to a better Internet. In *Proceedings of ReArch'08*, Madrid, Spain, December 2008. Co-located with ACM CoNEXT 2008.

15. Hoang T. Dinh, Chonho Lee, Dusit Niyato, and Ping Wang. A survey of mobile cloud computing: Architecture, applications, and approaches. *Wirel. Commun. Mob. Comput.*, 13(18):1587–1611, 2013.

16. Flavio Esposito, Francesco Chiti, and Romano Fantacci. Voice and data traffic classes management for heterogeneous networks enabled by mobile phones. In *INFOCOM*, pp. 27–36. Barcelona, Spain, April 2006.

17. Flavio Esposito, Francesco Chiti, Romano Fantacci, Junzhao Sun, and Simo Hosio. Agent based adaptive management of non-homogeneous connectivity resources. In *IEEE International Conference on Communications (ICC)*, pp. 82–88. Istanbul, Turkey, June 2006.

18. Flavio Esposito, Francesco Chiti, Junzhao Sun, Simo Hosio, and Romano Fantacci. Non-homogeneous connectivity management for GPRS and Bluetooth enabled networks. In *International Conference on Mobile and Ubiquitous Multimedia*, MUM, p. 17. Christchurch, New Zealand, December 2006.

19. Flavio Esposito, Donato Di Paola, and Ibrahim Matta. On distributed virtual network embedding with guarantees. *IEEE/ACM Trans. Netw.*, 24:569–582, 2016.

20. Flavio Esposito and Ibrahim Matta. PreDA: Predicate routing for DTN architectures over MANET. In *GLOBECOM IEEE Global Communication Conference*, p. 74. San Francisco, CA. September 2009.

21. Flavio Esposito, Ibrahim Matta, and Vatche Ishakian. Slice embedding solutions for distributed service architectures. *ACM Comput. Surv.* 46(1):1–29, 2014.

22. Flavio Esposito, Ibrahim Matta, and Yuefeng Wang. VINEA: An architecture for virtual network embedding policy programmability. *IEEE Tran. Parallel Distrib. Syst.*, (1), 2016.

23. Flavio Esposito, Yuefeng Wang, Ibrahim Matta, and John Day. Dynamic layer instantiation as a service. In *Demo at USENIX NSDI*, Lombard, IL, April 2013.

24. eXpressive Internet Architecture. 2010. http://cs.cmu.edu/x̄ia/

25. Kevin Fall. A delay-tolerant network architecture for challenged Internets. In *Proceedings of the 2003 Conference on Applications, Technologies, Architectures, and Protocols for Computer Communications*, SIG-COMM '03, pp. 27–34. New York: ACM, 2003.

26. Nick Feamster, Jennifer Rexford, and Ellen Zegura. The road to SDN: An intellectual history of programmable networks. *SIGCOMM Comput. Commun. Rev.*, 44(2):87–98, 2014.

27. Niroshinie Fernando, Seng W. Loke, and Wenny Rahayu. Mobile cloud computing. *Future Gener. Comput. Syst.*, 29(1):84–106, 2013.

28. Ines Houidi, Wajdi Louati, Walid Ben Ameur, and Djamal Zeghlache. Virtual network provisioning across multiple substrate networks. *Comput. Netw.*, 55(4):1011–1023, 2011.

29. Damien Saucez, Luigi Iannone, and Olivier Bonaventure. OpenLISP: An open source implementation of the locator/ID separation protocol. In *IEEE INFOCOM*, demo paper, Rio de Janeiro, Brazil, April 2009.

30. Institute of Electrical and Electronics Engineers. The 2nd IEEE International Workshop on Wireless Communications for High Speed Railways (HSRCom2016), Nanjing, China, May 2016. http://maxxx437.github.io/hsrcom2016/.

31. InterPlanetary Networking. http://www.ipnsig.org/.

32. Iperf Linux tool. https://iperf.fr/.

33. Iptables. Netfilter project. 2016. http://ipset.netfilter.org/iptables.man.html

34. Vatche Ishakian, Joseph Akinwumi, Flavio Esposito, and Ibrahim Matta. On supporting mobility and multihoming in recursive Internet architectures. *Comput. Commun.*, 35(13):1561–1573, 2012.

35. Prateek Joshi and C.-C Jay Kuo. Security and privacy in online social networks: A survey. In *2011 IEEE International Conference on Multimedia and Expo (ICME)*, pp. 1–6, July 2011.

36. Christoforos Kachris, Georgios Ch. Sirakoulis, and Dimitrios Soudris. Network function virtualization based on FPGAs: A framework for all-programmable network devices. In *CoRR*, abstract 1406.0309, 2014.

37. Eddie Kohler, Robert Morris, Benjie Chen, John Jannotti, and M. Frans Kaashoe. The Click modular router. *ACM Trans. Comput. Syst.*, 18(3):263–297, 2000.

38. Bob Lantz, Brandon Heller, and Nick McKeown. A network in a laptop: rapid prototyping for software-defined networks. In *Proceedings of ACM SIGCOMM*, HotNets-IX, pp. 19:1–19:6, 2010.

39. Nancy A. Lynch. *Distributed Algorithms*. Morgan Kaufmann Series in Data Management Systems, 1st ed. Burlington, MA: Morgan Kaufmann, 1996.

40. Nick McKeown, Tom Anderson, Hari Balakrishnan, Guru M. Parulkar, Larry L. Peterson, Jennifer Rexford, Scott Shenker, and Jonathan S. Turner. OpenFlow: Enabling innovation in campus networks. *SIGCOMM Comput. Commun. Rev.*, 38(2):69–74, 2008.

41. Mohamed Nidhal Mejri, Jalel Ben-Othman, and Mohamed Hamdi. Survey on {VANET} security challenges and possible cryptographic solutions. *Veh. Commun.*, 1(2):53–66, 2014.

42. David Meyer. The Locator/Identifier Separation Protocol (LISP). Internet Protocol J., 11(1), 2008.

43. MobilityFirst. 2010. http://mobilityfirst.winlab.rutgers.edu/

44. Named Data Networking. 2010. http://named-data.net/

45. NEBULA. 2010. http://nebula-fia.org/

46. OpenStack. 2013. http://openstack.org/

47. Optimized Network Engineering Tool (OPNET).

48. Bruno Quoitin, Luigi Iannone, Cédric de Launois, and Olivier Bonaventure. Evaluating the benefits of the locator/identifier separation. In *Proceedings of 2nd ACM/IEEE International Workshop on Mobility in the Evolving Internet Architecture*, pp. 1–6. New York: ACM, 2007.

49. Maxim Raya and Jean-Pierre Hubaux. The security of vehicular ad hoc networks. In *Proceedings of the 3rd ACM Workshop on Security of Ad Hoc and Sensor Networks*, SASN '05, pp. 11–21. New York: ACM, 2005.

50. RINA Research Group. Recursive InterNetwork Architecture 2010. http://csr.bu.edu/rina/. see also http://pouzinsociety.org/

51. Timothy Roscoe, Steven Hand, Rebecca Isaacs, Richard Mortier, and Paul W. Jardetzky. Predicate routing; enabling controlled networking. *Comput. Commun. Rev.*, 33(1):65–70, 2003.

52. Vsevolod Salnikov, Renaud Lambiotte, Anastasios Noulas, and Cecilia Mascolo. Openstreetcab: Exploiting taxi mobility patterns in New York City to reduce commuter costs. In *CoRR*, abstract 1503.03021, 2015. Available at https://arxiv.org/abs/1503.03021.

53. Brandon Schlinker, Kyriakos Zarifis, Italo Cunha, Nick Feamster, Ethan Katz-Bassett, and Minlan Yu. MiniNExT. Available at https://github.com/USC-NSL/miniNExT.

54. Kuldeep Singh, Poonam Saini, Sudesh Rani, and Awadhesh Kumar Singh. Authentication and privacy preserving message transfer scheme for vehicular ad hoc networks (VANETs). In *Proceedings of the 12th ACM International Conference on Computing Frontiers*, CF '15, pp. 58:1–58:7. New York: ACM, 2015.

55. Anirudh Sivaraman, Suvinay Subramanian, Anurag Agrawal, Sharad Chole, Shang-Tse Chuang, Tom Edsall, Mohammad Alizadeh, Sachin Katti, Nick McKeown, and Hari Balakrishnan. Towards programmable packet scheduling. In *Proceedings of the 14th ACM Workshop on Hot Topics in Networks*, HotNets-XIV, pp. 23:1–23:7. New York: ACM, 2015.

56. The IRATI Group. http://irati.eu/.

57. U.S. National Science Foundation. The FIND initiative. 2009. http://www.nets-find.net/

58. Luis M. Vaquero and Luis Rodero-Merino. Finding your way in the fog: Towards a comprehensive definition of fog computing. *SIGCOMM Comput. Commun. Rev.*, 44(5):27–32, 2014.

59. Anna Maria Vegni and Valeria Loscrí. A survey on vehicular social networks. *IEEE Commun. Surv. Tutorials*, 17(4):2397–2419, 2015.

60. Wenye Wang and I.F.A. Yildiz. On the estimation of user mobility pattern for location tracking in wireless networks. In *IEEE Global Telecommunications Conference 2002*, GLOBECOM '02, Vol. 1, pp. 610–614, November 2002.

61. Yi Wang, Ioannis Avramopoulos, and Jennifer Rexford. Morpheus: Making routing programmable. In *Proceedings of the 2007 SIGCOMM Workshop on Internet Network Management*, INM '07, pp. 285–286. New York: ACM, 2007. Kyoto, Japan.

62. Yuefeng Wang, Flavio Esposito, Ibrahim Matta, and John Day. Recursive InterNetworking Architecture (RINA) Boston University prototype programming manual. Technical Report BUCS-TR-2013-013. Boston: Boston University, November 2013.

63. Yuefeng Wang, Flavio Esposito, Ibrahim Matta, and John Day. RINA: An architecture for policy-based dynamic service management. Technical Report BUCS-TR-2013-014. Boston: Boston University, November 2013.

64. Yuefeng Wang, Ibrahim Matta, Flavio Esposito, and John Day. Introducing ProtoRINA: A prototype for programming recursive-networking policies. *SIGCOMM Comput. Commun. Rev.*, 44(3):129–131, 2014.

65. Md Whaiduzzaman, Mehdi Sookhak, Abdullah Gani, and Rajkumar Buyya. A survey on vehicular cloud computing. *J. Netw. Comput. Appl.*, 40:325–344, 2014.

66. Koji Yamazaki, Takeshi Osaka, Sadayuki Yasuda, Shoko Ohteru, and Akihiko Miyazaki. Accelerating SDN/NFV with transparent offloading architecture. Presented as part of the Open Networking Summit 2014 (ONS 2014), Santa Clara, CA, 2014.

67. Minlan Yu, Yung Yi, Jennifer Rexford, and Mung Chiang. Rethinking virtual network embedding: substrate support for path splitting and migration. *SIGCOMM Comput. Commun. Rev.*, 38(2):17–29, 2008.

68. Elena Zheleva and Lise Getoor. Privacy in social networks: A survey. In Social Network Data Analytics, pp. 277–306. Berlin: Springer, 2011.

69. Yu Zheng, Licia Capra, Ouri Wolfson, and Hai Yang. Urban computing: Concepts, methodologies, and applications. *ACM Trans. Intell. Syst. Technol.*, 5(3), 2014.

70. Yaping Zhu, Rui Zhang-Shen, Sampath Rangarajan, and Jennifer Rexford. In Cabernet: Connectivity architecture for better network services. In *CoNEXT*, pp. 64:1–64:6. New York: ACM, 2008.

APPLICATIONS AND SECURITY IN VSNs

Chapter 6

Socially Inspired Dissemination

Felipe D. Cunha, Guilherme Maia, and Antonio A. F. Loureiro
Federal University of Minas Gerais, Minas Gerais, Brazil

Leandro Villas
University of Campinas, São Paulo, Brazil

Aline Carneiro Viana
INRIA, Saclay, France

Raquel Mini
Pontifical University of Minas Gerias, Minas Gerais, Brazil

Contents

6.1 Introduction

Vehicular ad hoc networks (VANETs) are a special type of mobile ad hoc networks in which vehicles have processing and wireless communication capabilities. Usually, these vehicles exchange information among themselves through multihop communication. However, they can also communicate with special nodes placed near the roads, which are called roadside units (RSUs). The idea is that vehicles are able to establish communication under different environments, such as urban centers and highways, in order to cooperatively increase road safety and efficiency and provide entertainment to passengers [3].

In these networks, sending messages from a source to all vehicles located inside a geographic region will be very common. Such activity is known as data dissemination, and it is a required service for many applications. Data dissemination solutions must consider two important challenges. The first one, known as the *broadcast storm problem*, happens when a group of vehicles close to one another starts to transmit data messages at the same time, leading to a high number of message collisions and severe contention at the link layer [16, 18]. The second one, known as the *intermittently connected network problem*, happens in scenarios with low traffic densities, such as daybreak, holidays, and rural areas, in which the number of vehicles is not enough to disseminate data messages using direct multihop communication [15, 24].

A factor that contributes to the emergence of these problems is the driver's routine. Usually, people have similar behavior, which increases the likelihood of going to the same places, at the same time. Moreover, while moving around, drivers are susceptible to speed limits, traffic lights, obstacles, and so forth. Therefore, it is reasonable to assume that these factors combined lead to microscopic and macroscopic traffic density variations. We argue that a better understanding of these routines and their impact on the overall traffic conditions is fundamental in designing better communication protocols for VANETs, and in particular, in the context of this work, for tackling the two aforementioned problems faced by data dissemination solutions.

There is vast research that investigates the social aspects inherent to VANETs [4, 9, 12–14, 21]. In summary, they show that there are social properties encoded in vehicular networks. With this in mind, we leverage these social aspects to design a *socially inspired broadcast data dissemination* for VANETs. In our solution, we explore two approaches to performing data dissemination in VANETs: a delay-based solution and a probabilistic-based solution. In the delay-based solution, we use two social metrics to determine when vehicles should rebroadcast data messages. Differently, in the probabilistic-based solution, we use three social metrics to determine which vehicles should rebroadcast data messages. With these solutions, our goal is to increase the delivery ratio, decreasing the overall network overhead, independently of the road traffic conditions. Simulation results show that when compared with three related solutions—UV-CAST [24], ABSM [15], and AID [1]—under a Manhattan grid and a city scenario based on a realistic mobility dataset, our solutions possess a higher delivery ratio, decrease both the number of collisions and the total number of data messages transmitted, and have an acceptable delay.

The remainder of this chapter is organized as follows. Section 6.2 presents the definition for data dissemination and an overview of recent related work. Section 6.3 discusses how social metrics can help to improve communication protocols for VANETs. Section 6.4 presents our two socially inspired solutions. Section 6.5 describes the simulation scenarios and metrics and presents the simulation results. Finally, Section 6.6 presents our conclusions.

6.2 Data Dissemination

Many are the vehicular applications that use the data dissemination process to deliver the message and provide their services. However, each application has specific requirements, which demands different strategies to promote data dissemination. In this section, we present more details about the data dissemination process (Section 6.2.1) and a state-of-the-art review (Section 6.2.2).

6.2.1 Definition

Data dissemination corresponds to the process in which a single source vehicle or RSU broadcasts data messages to all vehicles located inside a region of interest (ROI) through multihop communications, as illustrated in Figure 6.1. The ROI is defined by the application for which the messages must be disseminated. Moreover, in this work, for the sake of simplicity, we assume the ROI is defined as a circular region centered at the source. However, any kind of region may be employed, as long as a vehicle is able to determine whether it is inside such a region or not. The main goal of this process is to guarantee message delivery to all vehicles inside the ROI independently of the road traffic conditions. Moreover, according to application requirements, the data dissemination process can be aware of other metrics, such as delays, collisions, and overhead. Therefore, the solution must be able to operate under both dense (Figure 6.1a) and sparse (Figure 6.1b) scenarios, which requires tackling the broadcast storm and intermittently connected network problems.

6.2.2 State of the Art

The study to look at computer networks as social networks has been increased, because these are networks that link people, organizations, and knowledge. In literature, we can find a lot of research that use properties and characteristics to improve the performance of protocols and services in computer networks [6, 8, 10, 11, 22]. Specific to vehicular networks, in [4, 14], the authors present a discussion about social features and how human life can impact vehicular networks.

In the traditional ad hoc networks, SimBet [6] uses centrality metrics and social similarity to define the probability of a node contacting the destination node. The goal of this protocol is to find bridge nodes able to connect two communities. Thus, it is possible to increase the broadcast coverage. In a different way from our solution, in [8], the authors present a data dissemination protocol. This protocol defines the retransmission probability for a node to be inversely proportional to its number of neighbors. However, the focus of this protocol is to perform data dissemination in high-density environments. In [22], the authors use small-world concepts to design a wireless

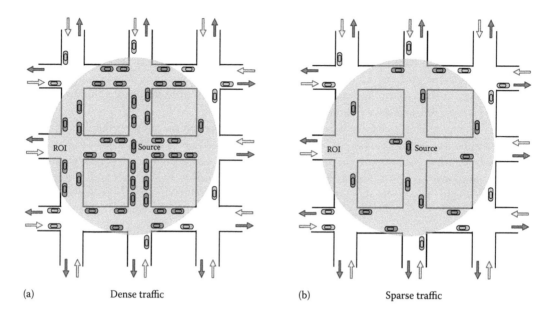

(a)   Dense traffic   (b)   Sparse traffic

Figure 6.1   Data dissemination to a group of vehicles under both dense and sparse traffic scenarios. (From F. D. Cunha et al., in *Proceedings of the 17th ACM International Conference on Modeling, Analysis and Simulation of Wireless and Mobile Systems* [MSWiM '14], pp. 81–85, New York: ACM, 2014.)

mesh network. Due to the features and shortcomings faced by wireless mesh networks, such as throughput degradation and poor capacity scaling, the authors outline strategies to create shortcuts between router nodes, aiming to decrease communication delay.

However, in vehicular networks, we have some peculiarities, different from those of the traditional ad hoc networks. Thus, the traditional solutions proposed for mobile ad hoc networks cannot be applied directly to vehicular networks. In VANETs, vehicles portray higher speeds, which lead to a lot of changes in the network topology and very fast contacts among the vehicles. Furthermore, the traffic condition depend on the region (rural area, highway, or downtown), the time of day (rush hour or daybreak), and the weekday (workday, weekend, or holiday).

Many are the solutions proposed specifically for vehicular networks to perform data dissemination [1, 15, 16, 18, 23, 24]. For instance, in [18], the authors present DV-CAST, which aims to solve both problems (broadcast storm and network partitions). DV-CAST uses periodic beacon messages to build the local topology (one-hop neighbors) that is used to rebroadcast a message. DV-CAST performs data dissemination in both dense and sparse networks. For this, during data dissemination, the receiver applies the broadcast suppression algorithm if the local topology is well connected; otherwise, it uses the store-carry-forward mechanism as a solution in a sparsely connected neighborhood. However, in scenarios with high mobility, the DV-CAST performance depends on the beacon frequency, whose optimal value is difficult to establish. In addition, DV-CAST focuses only on highway topologies.

The Simple and Robust Dissemination (SRD) [16] solution was conceived to operate under dense and sparse vehicular networks and is an improvement over DV-CAST. Similar to DV-CAST, SRD relies exclusively on local one-hop neighbor information, and it does not employ any special infrastructure. Among the main improvements, when compared with DV-CAST, SDR proposes an *optimized slotted-1-persistence* broadcast suppression technique. Under this scheme, vehicles have different priorities to rebroadcast according to their moving direction. SRD prevents the broadcast storm problem in dense networks, and it also deals with disconnected networks by relying on the store-carry-forward communication model. However, SRD performs directional data dissemination in highway environments.

In [15], Ros et al. proposed the ABSM. This is a solution to perform data dissemination in VANETs with varying road traffic conditions. The main idea of ABSM is to use the minimum connected dominating set (MCDS) concept to determine the vehicles that will forward the message during the dissemination process. Then 100% coverage is guaranteed with a low overhead. However, determining the MCDS is a NP-hard problem. For that reason, the authors define a distributed heuristic to determine whether a vehicle belongs to the MCDS. Thus, the vehicles in the MCDS are assigned a lower waiting delay to rebroadcast the message. Moreover, aiming to guarantee the message delivery in sparse scenarios, ABSM relies on periodic beacons to advertise to neighbors the presence of new vehicles in the vicinity. Then, when a vehicle receives a beacon from a neighbor and it does not acknowledge the receipt of a message, the vehicle forwards the message to the neighbor. However, this is a delay-based solution and is not applied to scenarios with delay-sensitive applications.

Another solution proposed for VANETs is AID [1]. This is a decentralized and adaptive solution for information dissemination in VANETs, where the vehicles deciding whether to forward the message depends on the probability. This probability is set up based on the number of times that the vehicles receive the same message in a given period of time. In scenarios with dense traffic, several vehicles might decide to drop the message since it has already been forwarded by several vehicles, reducing the broadcast storm problem. However, this solution does not deal with the problem of intermittent connections on the network.

Taking into account the solution designs for both scenarios, UV-CAST [24] is proposed to perform data dissemination in different traffic conditions. Thus, the vehicles work in two states: broadcast suppression or store-carry-forward. When a vehicle receives a data message for the first time, it initially checks whether it is a border vehicle. Border vehicles are the ones that are at the edge of a connected component. Thus, UV-CAST assumes these vehicles have a higher probability of meeting new neighbors. If the vehicle verifies it is a border vehicle, then it stores the message and carries it around until an encounter with a new neighbor is made. On the other hand, if the vehicle is not a border vehicle, it executes a broadcast suppression algorithm to rebroadcast the message.

Aiming to tackle the two common problems in dissemination, in [23], Villas et al. propose a new solution to perform data dissemination in VANETs, considering both scenarios: dense and sparse networks. DRIVE relies exclusively on local one-hop neighbor information to deliver messages in these scenarios. Based on a preference zone and the distance of the transmitter, it defines the delay on retransmission. Also, the solution employs implicit acknowledgments to guarantee robustness in message delivery under sparse scenarios. In Table 6.1, we present a comparison of these solutions, summarizing the main features of each one.

6.3 Social Metrics Applied

The use of social metrics to improve the performance of protocols and services in ad hoc networks has received much attention by the research community. In the literature, we can find many works that describe metrics evaluation, characterization, and new services that use these metrics [4, 9–11, 14, 21]. In particular, for VANETs, protocols and services can improve considerably when we consider drivers' behavior and routines. With such knowledge, for instance, it is possible to better understand daily traffic evolution and adapt communication protocols accordingly.

Table 6.1 Comparison of Data Dissemination Solutions for VANETs

Dissemination Solution	Forwarding Strategy	Architecture	Scenario	Assumptions
DV-CAST [18]	Position based, store-carry-forward	V2V	Highway	GPS and neighbor's position required
SRD [16]	Position and distance based, store-carry-forward	V2V	Highway	GPS and neighbor's position required
ABSM [15]	Position and delay based	V2V	Urban and highway	GPS receiver required
AID [1]	Statistical based	V2V	Urban	–
UV-CAST [24]	Position and store-carry-forward	V2V	Urban	GPS and neighbor's position required
DRIVE [23]	Position, distance, and timer based	V2V	Urban and highway	GPS receiver and map required

Below, we describe a group of social metrics and the influence of each one in the data communication in VANETs. Taking into account the vehicle's mobility, we classify these metrics into macroscopic and microscopic metrics [19]. The macroscopic metrics represent measures of the network global state, which can portray the general behavior of all vehicles. The microscopic metrics define individual values for the vehicle, representing the behavior of a unique vehicle. For macroscopic metrics, we choose the distance, diameter, density, and edge persistence, and as microscopic metrics, we select node degree, closeness centrality, cluster coefficient, and topological overlap.

6.3.1 Macroscopic Metrics

Distance: This metric portrays the average number of hops for a vehicle to reach another. In many cases, to reduce the delay in which messages are delivered to intended recipients, it may be interesting to use the shortest routes. Moreover, vehicles that are close to one another may share a common interest, which may be helpful in directing data dissemination flows.

Diameter: In accordance with the distance, this metric captures the major distance between two vehicles in the graph. When analyzing the network topology, a great diameter may indicate a high communication cost to reach all vehicles, due to the high number of hops.

Density: This metric represents how dense is the network, that is, the average number of connections in the network. In the context of VANETs, urban regions can have higher densities than rural areas. This is particularly the case for downtown regions, where the road traffic is very dense. In possession of such knowledge, data dissemination solutions can adapt to the perceived traffic condition in order to determine whether vehicles should operate under a broadcast suppression or a store-carry-forward state.

Edge persistence: This metric represents the persistence of an encounter between two vehicles. In some cases, the encounter happens in the same region, which indicates similar routines between two vehicles. For various vehicular applications, to reach a given destination, vehicles can use these persistent connections as a backbone for data forwarding.

6.3.2 Microscopic Metrics

Node degree: This metric presents information about the local neighborhood of a node. In multihop data communication solutions, neighbors are used as relays to deliver data to intended recipients. For the particular case of data dissemination protocols, a vehicle with a high degree has greater coverage; that is, it can deliver messages to more vehicles at once.

Closeness centrality: This metric measures the centrality of the vehicle according to its distance from others vehicles in the network. In VANETs, vehicles with a high closeness centrality are closer to the remaining vehicles in the network. Therefore, the dissemination can be faster when choosing these vehicles as data forwarders.

Clustering coefficient: This metric evaluates how close are the neighbors of a vehicle. Commonly, vehicles that belong to a group have similar features and behaviors, and information can be useful for the whole cluster. Thus, taking into account the wireless broadcast advantage, with just one transmission from a group member, it is possible to reach the whole cluster, reducing the overall number of transmissions.

Topological overlap: This metric measures the percentage of neighbors that are shared among two or more nodes. In a vehicular network, vehicles that share interests and have similar behaviors tend to group together. Normally, they have a high percentage of shared neighbors. Thus, in a dissemination process, the vehicle that has a small topological value should be used to reach different vehicles, increasing the coverage ratio.

6.4 Socially Inspired Dissemination Solutions

Many are the strategies defined to perform data dissemination in vehicular networks. These strategies must work to deliver the message to all intended vehicles in different road traffic conditions. In this section, we present two data dissemination solutions, which employ two different strategies: a delay-based solution (Section 6.4.1) and a probabilistic-based solution (Section 6.4.2).

6.4.1 Delay-Based Solution

The main idea of a delay-based solution is to determine which vehicles should rebroadcast a received data message and when they should do so. However, this process should happen for both dense and sparse road traffic scenarios. Algorithms 6.1 and 6.2 define the main steps of this process and how the vehicle calculates a waiting delay to rebroadcast the message.

For both algorithms, we assume that vehicles store and carry each received data message for the whole period in which they are inside the ROI and the time-to-live for the message has not expired. Moreover, they are equipped with a global positioning system (GPS), or they can infer their positions through other means. Each vehicle periodically exchanges beacons with its neighbors. These beacons contain context information about the vehicle, for instance, the position and number of neighbors (node degree). Furthermore, each beacon contains the IDs of the data messages that have been received and are being carried by the vehicle. Notice that embedding the IDs of received data messages into beacons works as an implicit acknowledgment mechanism. Therefore, when a vehicle receives a beacon from a neighbor, it is able to verify whether it possesses any data message that has not been received by this neighbor, and then forward it accordingly.

In the following two sections, we thoroughly describe each algorithm. Thereafter, we show how the clustering coefficient, the node degree, and the combination of both metrics can be used to turn our data dissemination solution into a delay-based solution. Initially, we show how to estimate the clustering coefficient using only one-hop neighbor information, and how to use it in the waiting delay computation. We then turn our attention to the node degree, which can be easily obtained through beacons. Finally, we also show how to calculate the waiting delay using a combination of both metrics.

6.4.1.1 Broadcast Suppression

Under dense road traffic conditions, when a vehicle receives a data message, it must carefully decide whether to rebroadcast it, and when to rebroadcast it in order to avoid redundant retransmissions and, consequently, the broadcast storm problem. Algorithm 6.1 shows how a vehicle proceeds when it receives a data message m.

Initially, the vehicle verifies whether it has left the ROI or the time-to-live for the message m has expired. In such a case, the vehicle discards m (lines 1–3). Otherwise, the vehicle checks whether m is a duplicate (line 4). If it is not a duplicate, then the vehicle stores m in the list of received messages that are still valid. Furthermore, it will insert the ID of m into subsequent beacons, until the vehicle leaves the ROI or m expires (lines 5 and 6). The next and most important step is to calculate the waiting delay t to rebroadcast m (line 7). In Algorithm 6.1, we omitted how this delay is calculated, because it will depend on the social metric employed. For now, it is enough to know that such a delay is a value in the interval $[0, T_{max}]$, where T_{max} is a configured parameter. After calculating the waiting delay, the vehicle uses it to schedule a rebroadcast for m (line 8). Notice that while the vehicle is scheduled to rebroadcast m, if it receives a duplicate, then it cancels the rebroadcast (lines 10 and 11), thus avoiding a possible redundant retransmission. However, when the waiting delay expires and the vehicle has not received any duplicate, it rebroadcasts m (lines 15 and 16).

Algorithm 6.1: Broadcast Suppression Algorithm

Require: Data message *m* received from neighbor *s*

1: **if** Vehicle is outside the region of interest specified in *m* or the time-to-live of *m* expired **then**
2:   discard *m*;
3: **end if**
4: **if** *m* is not a duplicate **then**
5:   add message to the list of received messages;
6:   insert *m* ID in subsequent beacons;
7:   $t \leftarrow$ calculateWaitingDelay();
8:   schedule *rebroadcast_timer* for *m* to fire up at *currentTime* + *t*;
9: **else**
10:   **if** *rebroadcast_timer* for *m* is scheduled **then**
11:     cancel *rebroadcast_timer* for *m*;
12:   **end if**
13: **end if**
14: **Event:** scheduled *rebroadcast_timer* for *m* expires
15: Rebroadcast *m*;

6.4.1.2 Store-Carry-Forward

On the other hand, when the road traffic is sparse and the network is partitioned, vehicles must hold received data messages and use their mobility capabilities to carry the messages to different parts of the ROI. Moreover, they must be able to determine whether a vehicle has already received a data message. For the former issue, vehicles rely on the store-carry-forward communication model. For the latter, beacons are used as an implicit acknowledgment mechanism. Algorithm 6.2 shows how our proposed solution delivers data messages even when the network is intermittently connected.

Algorithm 6.2: Store-Carry-Forward Algorithm

Require: Beacon *b* received from neighbor *s*

1: **for all** message *m* in the list of received messages **do**
2:   **if** *m* is not acknowledged in *b* **then**
3:     $t \leftarrow$ calculateWaitingDelay();
4:     schedule *rebroadcast_timer* for *m* to fire up at *currentTime* + *t*;
5:   **end if**
6: **end for**
7: **Event:** data message *m* received from neighbor *s*
8: **if** *m* is a duplicate **then**
9:   **if** *rebroadcast_timer* for *m* is scheduled **then**
10:     cancel *rebroadcast_timer* for *m*;
11:   **end if**
12: **end if**
13: **Event:** scheduled *rebroadcast_timer* for *m* expires
14: Rebroadcast *m*;

When a vehicle receives a beacon *b* from a neighbor *s*, it verifies whether there is a data message that has not been acknowledge by *s* in *b* (lines 1 and 2). For that, the vehicle looks into its list of received messages and compares their IDs with the IDs contained in *b*. If the vehicle finds any message *m* that has not been acknowledged, then it calculates a waiting delay *t* to rebroadcast *m* (line 3). Once again, such a delay will depend on the social metric employed. After calculating the waiting delay, the vehicle schedules to rebroadcast *m* with delay *t* (line 4). As in the broadcast suppression algorithm, while the vehicle is scheduled to rebroadcast *m*, if it receives a duplicate, then it cancels the rebroadcast (lines 9–12), thus avoiding a possible redundant retransmission. However, when the waiting delay expires and the vehicle has not received any duplicate, it rebroadcasts *m* (lines 15 and 16).

By using these two algorithms in conjunction, our proposed solution is able to tackle both the broadcast storm and the intermittently connected network problems. Moreover, it is worth noticing that a vehicle does not need to be aware of the current road traffic conditions, that is, whether the network is dense or sparse. In either case, the vehicle always tries to avoid redundant retransmissions and increase the message delivery capability to intended recipients.

6.4.1.3 Clustering Coefficient Solution

The clustering coefficient for a vehicle *v* is the number of connections between neighbors of *v* divided by the total number of possible connections between neighbors of *v* [25]. Therefore, to accurately calculate the clustering coefficient for vehicle *v*, it is necessary to know the two-hop neighborhood knowledge of *v*. Given that VANETs are extremely dynamic networks and obtaining such knowledge can be cumbersome, here we use position information to estimate the clustering coefficient, in particular to determine whether two neighbors of a vehicle are connected. As already stated, each vehicle knows the position of each neighbor due to received beacons. Therefore, to verify whether two neighbors are connected, vehicle *v* must only check whether the distance between these two neighbors is below the estimated communication range. Thereafter, *v* is able to calculate its estimated clustering coefficient.

In possession of its own estimated clustering coefficient, a vehicle *v* is able to calculate its waiting delay to rebroadcast. According to an analysis of the estimated clustering coefficient with respect to the vehicle density (see Figures 6.3 and 6.4), for lower densities, the clustering coefficient is also low, but the variability is high. On the other hand, when the density is high, so, too, is the value of the estimated clustering coefficient, but the variability is low. For our purposes, the greater the variability, the better. Otherwise, we risk assigning the same or similar waiting delay to all vehicles. Therefore, for this first proposal, we give a higher priority to rebroadcast for vehicles that have a low estimated clustering coefficient. In other words, the lower the estimated clustering coefficient, the lower the waiting delay. Equation 6.1 shows how the waiting delay is calculated using this metric, where the value for *estimatedCC* ranges in the interval [0, 1].

$$t = T_{max} \times estimatedCC \tag{6.1}$$

6.4.1.4 Node Degree Solution

When we look into the analysis of the node degree (see Figures 6.3 and 6.4), we can see that when the vehicle density is low, the degree and its variability are also low. However, when the density increases, both the degree and its variability increase. Therefore, in the proposal based on the node degree, we use an opposite approach. That is, the higher the degree of a vehicle at a given neighborhood, the higher its priority to rebroadcast the message, that is, the lower the waiting delay. Each vehicle will know its max neighbor degree due to the *degree* information in the received beacons. Equation 6.2 shows how the waiting delay is calculated using this approach. Here, *degree* is the degree of the vehicle that is calculating the waiting delay and *maxDegree* is the maximum between *degree* and the highest degree among all neighbors of the vehicle.

$$t = T_{max} \times \left(1 - \left(\frac{degree}{maxDegree} \right) \right) \tag{6.2}$$

6.4.1.5 Joint Solution

Here, we also propose a joint solution, that is, one that uses both the estimated clustering coefficient and the node degree. The idea is that, assuming that a single metric may not be adequate for all traffic density scenarios, a combination of the two may produce better results. Equation 6.3 shows how the waiting delay can be calculated using this joint solution. As can be observed, each metric contributes to a fraction of the total waiting delay, which is controlled by the factors α and β. To balance the delay equation, we assume that $\alpha = \beta = 0.5$.

$$t = t_{cc} + t_{degree} \tag{6.3}$$

$$t_{cc} = \alpha T_{max} \times estimatedCC \tag{6.4}$$

$$t_{degree} = \beta T_{max} \times \left(1 - \left(\frac{degree}{maxDegree} \right) \right) \tag{6.5}$$

6.4.2 Probabilistic-Based Solution

Another strategy to perform data dissemination is using a probabilistic-based approach. The main idea is to probabilistically determine which vehicles should rebroadcast the data message received. This process is made by deriving a probability for each node based on its social metrics. We assume that in this solution, the advertisement beacon mechanism works in the same way as in the delay-based solution (see Section 6.4.1). Algorithms 6.3 and 6.4 define the main steps of the probability computation process. In the following two sections, we thoroughly describe each algorithm. Next, we describe the two ways to define the probability. First, the probability is defined based on the joint metrics node degree and clustering coefficient. Second, the probability is defined considering the metric topological overlap.

6.4.2.1 Broadcast Suppression

Aiming to deal with the broadcast storm problem, when a vehicle receives a data message, it must decide whether to rebroadcast it, avoiding redundant retransmissions. Algorithm 6.3 shows the steps that are executed when a vehicle receives a data message *m*.

At the beginning, the vehicle verifies whether it is outside the ROI or the time-to-live of the message *m* has expired. In these cases, the message *m* is discarded (lines 1–3). However, it verifies whether the message is duplicated (line 4). If it is not a duplicate, then the vehicle stores *m* in the list of received messages that are still valid. Then, it inserts the ID of *m* into the next beacons while it is inside the ROI or the time-to-live of *m* has not expired (lines 5 and 6). In the next step, the vehicle calculates the probability to rebroadcast *m* (line 7), which depends on the metric chosen. Thus, it generates a random value between 0 and 1 and rebroadcasts the message *m* if the random value is less than or equal to *p* (lines 8–10). If the vehicle receives the same message *m* while the rebroadcast is scheduled, it cancels the rebroadcast (lines 13 and 14), which contributes to reduce the number of redundant retransmissions and collisions (lines 18 and 19). Then, following the algorithm, the vehicle rebroadcasts the message when the scheduled timer expires.

Algorithm 6.3: Broadcast Suppression Algorithm

Require: Data message m received from neighbor s

1: **if** vehicle is outside the region of interest specified in m or the time-to-live of m expired **then**
2:     discard m;
3: **end if**
4: **if** m is not a duplicate **then**
5:     add message to the list of received messages;
6:     insert m ID in subsequent beacons;
7:     $p \leftarrow$ calculateTransmissionProbability();
8:     *random* $\leftarrow$ choose a random number between 0 and 1;
9:     **if** *random* $\leq p$ **then**
10:        schedule the *rebroadcast* for m;
11:    **end if**
12: **else**
13:    **if** *rebroadcast* for m is scheduled **then**
14:       cancel *rebroadcast* for m;
15:    **end if**
16: **end if**
17: **Event:** scheduled *rebroadcast* for m expires
18: Rebroadcast m;

6.4.2.2 Store-Carry-Forward

In the other scenario, when the traffic density is low and the network presents partitions, the vehicles need to hold the received data message in order to carry it to different parts of the ROI. In the same way as for the delay solution, the vehicles need to determine whether a vehicle has already received a data message. Thus, they implement a store-carry-forward communication model. Algorithm 6.4 presents the steps that are executed when the network is intermittently connected.

When a vehicle receives a beacon b from a neighbor s, it verifies whether there is some data message not acknowledged by neighbor s in beacon b (lines 1 and 2). For each message without an acknowledgment, the vehicle calculates the probability to rebroadcast and schedules the rebroadcast (line 3). Again, this probability will be defined according to the metric chosen. Following, it chooses a random value between 0 and 1. If this *random* value is less than or equal to the probability p, the vehicle will rebroadcast the message (lines 4–8). If during this process the vehicle receives any message m from some neighbor, it cancels the rebroadcast of m (lines 12–15), which contributes to reducing the redundant retransmissions. Otherwise, it rebroadcasts message m (lines 18 and 19).

6.4.2.3 Node Degree and Clustering Coefficient Solution

The probability definition given by the joint solution considers the metrics clustering coefficient and node degree. The focus is to favor the vehicles that have a higher degree and higher clustering coefficient. Then, the probability is defined in the following way: $prob = \alpha prob1 + \beta prob2$. Each metric will contribute to generate the final value. Thus, in order to balance in an equal way, we choose the value 0.5 for the variables α and β.

The value of *prob1* is given by the metric clustering coefficient. In Figure 6.2a, we can see the probability for different values of this metric. As we can see, considering the clustering coefficient (Figure 6.2a), the retransmission probability value *prob1* is equal to the clustering coefficient,

---

Algorithm 6.4: Store-Carry-Forward Algorithm

Require: Beacon *b* received from neighbor *s*
1: **for all** message *m* in the list of received messages **do**
2:    **if** *m* is not acknowledged in *b* **then**
3:       *p* ← calculateTransmissionProbability();
4:       *random* ← choose a random number between 0 and 1;
5:       **if** *random* ≤ *p* **then**
6:          schedule the *rebroadcast* for *m*;
7:       **end if**
8:    **end if**
9: **end for**
10: **Event:** data message *m* received from neighbor *s*
11: **if** *m* is a duplicate **then**
12:    **if** *rebroadcast* for *m* is scheduled **then**
13:       cancel *rebroadcast* for *m*;
14:    **end if**
15: **end if**
16: **Event:** scheduled *rebroadcast* for *m* expires
17: Rebroadcast *m*;

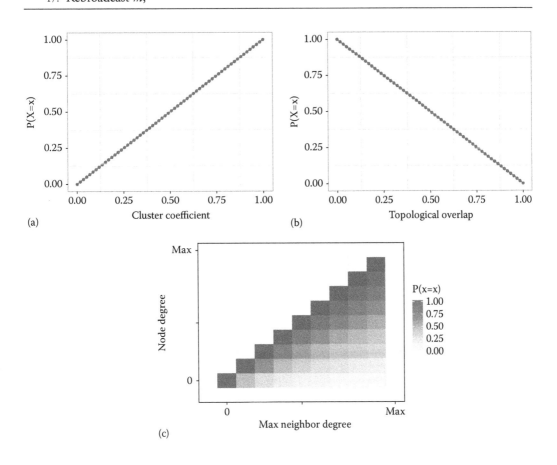

Figure 6.2 **Forwarding probability for each metric.**

where, in the same approach as for the delay-based solutions, this value ranges in the interval [0,1].

The computation of the probability *prob2* is given by Equation 6.6, where the probability is defined by the ratio of the node degree and the higher degree value in the neighborhood. Thus, to perform this computation, each vehicle sends its degree value into the beacons. Figure 6.2c shows the probability for different values of degree and max neighbor degree. It is possible to see that the *prob2* favors the vehicles that present the max degree to rebroadcast the data message.

$$prob2 = \frac{degree}{maxDegree} \tag{6.6}$$

6.4.2.4 Topological Overlap Solution

The topological overlap of a vehicle v defines the ratio of neighbors that v shares with its neighbors, that is, the fraction of common neighbors among the vehicles. Normally, these vehicles are connected with one another, forming a group of vehicles that share interests. Equation 6.7 presents the computation of the topological overlap [7].

$$to_{(i,j)} = \frac{|\{k|(i,k) \in E_t\} \cap \{k|(j,k) \in E_t\}|}{|\{k|(i,k) \in E_t\} \cup \{k|(j,k) \in E_t\}|} \tag{6.7}$$

In this way, aiming to compute this metric, each vehicle will send its ID and its neighbors' information together in the message beacons. The key idea in choosing this metric is to increase the data dissemination coverage. As we can see in Figure 6.2b, the smaller the value of the topological overlap of a vehicle, the greater will be its probability to rebroadcast. Thus, the probability is defined by the equation $prob = 1 - Topology$.

6.5 Performance Evaluation

For the evaluation of our proposed approaches, we executed a group of simulations using the OMNeT++ 4.2.2 network simulator [20]. In all results, we compare our proposals to state-of-the art solutions found in the literature, where two are delay-based solutions (UV-CAST [24] and ABSM [15]) and one is probabilistic based (AID [1]). Furthermore, in the next results, we nominated the solutions as CC-Degree (joint solution) and Topology (topological overlap solution). Following, we describe the details of the simulations and the results. Section 6.5.1 provides details about the scenarios used in the simulations, Section 6.5.2 describes the default parameters, and Section 6.5.3 describes the metrics used in our evaluation. In Section 6.5.4, we present and discuss the results for the delay-based solutions, and in Section 6.5.5, we present the results for the probabilistic-based solutions.

6.5.1 Simulation Scenarios

In order to evaluate the performance of the solutions, we used two different scenarios: a scenario that favors the occurrence of the broadcast storm problem and another scenario that favors intermittent connections and network partitions.

6.5.1.1 Manhattan Scenario

This is a scenario with 10 evenly spaced double-lane streets in an area of 1 km². Also, we consider signal attenuation effects caused by buildings. For that, we assume that each block has an 80 m × 80 m obstacle, which represents high-rise buildings. In order to quantify the traffic evolution in this scenario, we vary the vehicle density from 20 to 500 vehicles/km^2. The road traffic simulation is performed by the Simulator of Urban MObility (SUMO 0.17.0) [2]. Moreover, we positioned the source vehicle at the center of the grid, and it generates 100 messages of 2048 bytes to be disseminated to the whole network. The data rate is set to 1.5 Mbit/s.

To better understand the Manhattan grid scenario, Figure 6.3 shows the estimated clustering coefficient and the node degree for the considered vehicle densities. In particular, Figure 6.3a shows the estimated cluster coefficient and its evolution. As we can see, the value of the estimated cluster coefficient under low densities is small, almost 40%. Moreover, it has a higher variability. This happens because, for lower densities, there are few vehicles in transit. With the growth of the density, the estimated cluster coefficient increases. This is due to the fact that under higher densities, the encounter is probability also higher, and the network will be more connected. Therefore, starting at 200 vehicles/km^2, the value for the estimated clustering coefficient has a constant behavior of about 75%. This can be explained by the fact that even if the density of the network increases, connections among vehicles are constrained by physical restrictions, such as road shapes and obstacles.

Figure 6.3b presents the node degree evolution. It is possible to observe how the node degree evolves over the density variation. As expected, with the increase of the density, the node degree

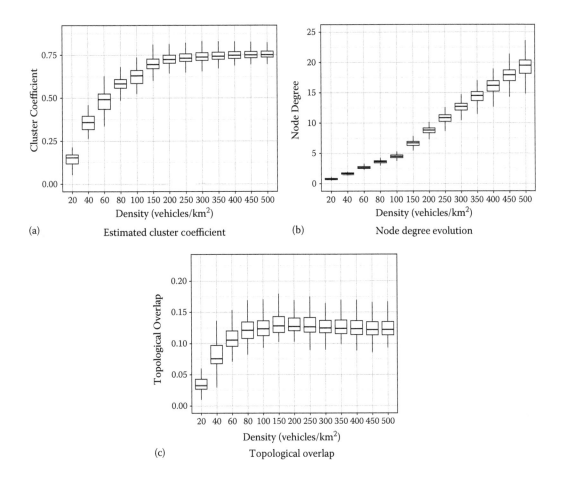

(a)   Estimated cluster coefficient

(b)   Node degree evolution

(c)   Topological overlap

Figure 6.3   **Manhattan metrics evolution.**

**Table 6.2 Vehicle Density during the
Time Evolution for the Cologne Scenario**

Time (a.m.)	Density (vehicles/km^2)
6:30	61
6:45	82
7:00	92
7:15	102
7:30	108

also increases. For instance, at 100 vehicles/km^2, the average node degree is 5, representing that, on average, a vehicle has five neighbors. We also can observe that at higher densities, the variability of the node degree is higher. In Figure 6.3c, we present the topological overlap evolution. We note that at lower density, the network presents low values to the topological metric, due to the fact that we have a sparser network. In higher densities, the topology overlap reaches a constant value, close to 15% of neighbors shared. In this case, we have a more connected network and a tendency to share more neighbors.

6.5.1.2 Cologne Trace

This is a realistic scenario represented by a 2-hour mobility dataset covering an area of 400 km^2 over the city of Cologne, Germany [19]. This trace has more details related to the physical infrastructure than does the Manhattan grid, for instance, traffic light signalization, different road types, and buildings, which makes it much more realistic. In order to consider different road traffic conditions, the data dissemination process is executed at different instants of the 2-hour dataset (6:30, 6:45, 7:00, 7:15, and 7:30 a.m.). To better understand the traffic evolution in this scenario and to facilitate comparisons with the Manhattan grid scenario, Table 6.2 shows the traffic density for each moment the dissemination is performed. As in the Manhattan grid scenario, a vehicle at the center of the network starts the dissemination. The vehicle generates 100 messages of 2048 bytes, at a data rate of 1.5 Mbit/s. The intended recipients for the dissemination are all vehicles inside a ROI with a radius of 2 km and centered at the source vehicle.

As for the Manhattan grid, Figure 6.4 shows the estimated cluster coefficient and node degree evolution for the Cologne scenario. For the estimated cluster coefficient (Figure 6.4a), we can observe a high variation for all considered times of the day. For instance, at 6:30 a.m., these values range from 0.15 to 1, where most vehicles have an estimated cluster coefficient between 0.6 and 0.9, which indicates a highly clustered network. Figure 6.4b shows the node degree evolution. Once again, we can observe a high variation, from 0 to 50 neighbors. These behaviors are expected due to traffic variations during rush hours. Moreover, in Figure 6.3c, we can see the evolution of the topological overlap. With the same behavior, we can observe a higher variation of this metric, from 0% to 75% of neighbors shared, and a great number of vehicles presenting 25% of neighbors shared. This happens because this scenario represents the beginning of the day where the traffic is sparse with some points of density traffic.

6.5.2 Simulation Parameters

We execute the simulation using the OMNeT++ network simulation and the framework Veins 2.1 [17], specific to vehicular networks. Our focus is to improve the quality of the following results and make them more realistic. This framework implements the IEEE 802.11p standard protocol stack for vehicle communication and an obstacle model for signal attenuation. Moreover, we set

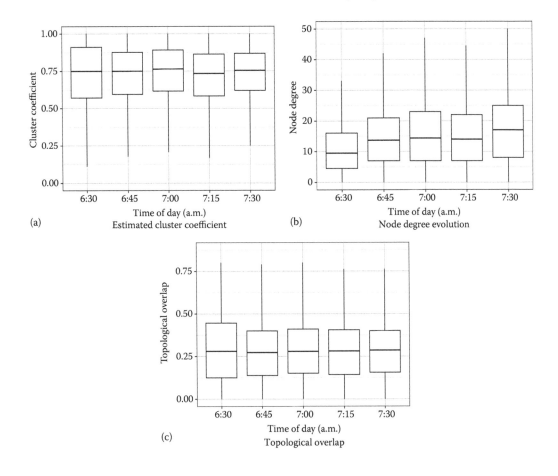

Figure 6.4   Cologne metrics evolution.

the bit rate at the media access control (MAC) layer to 18 Mbit/s and the transmission power to 0.98 mW. With these parameters and a two-ray ground propagation model, it is possible to reach a communication range of 200 m. Communication beacons are sent every 1 s. For all scenarios, we simulate r replications aiming to compute the confidence interval of 95% in the results. In Table 6.3, we summarize the main parameters used in the simulations.

6.5.3 Evaluated Metrics

Below, we present the metrics used to evaluate the performance of our solutions. The focus is to verify the coverage of the solution, the overhead induced by data messages, and the latency for different network density conditions.

- **Delivery ratio:** The percentage of data messages generated by the source that is delivered to intended recipients. The closer to 100% the delivery ratio is, the greater the reliability of data dissemination.

- **Total messages transmitted:** This metric computes the total number of data messages transmitted during the dissemination. It is an important metric to capture the number of redundant retransmissions, which may cause the broadcast storm problem.

Table 6.3   Default Simulation Parameters

Parameter	Value
Transmission power	0.98 mW
Transmission range	200 m
MAC bit rate	18 Mbit/s
Maximum waiting delay (T_{max})	500 ms
Beacon frequency	1 Hz
Data message size	2048 bytes
Number of data messages produced	100
Confidence interval	95%

■ **Collisions:** The average number of collisions per vehicle to perform the dissemination. It is desired for a solution to induce a low number of message collisions.

■ **Delay:** This is the average time for a data message to travel from the source vehicle to the intended recipients. Some applications in VANETs have hard delay requirements, such as alert services. For these types of services, messages must be disseminated as quickly as possible.

6.5.4 *Delay-Based Results*

Figure 6.5 shows the results for the Manhattan grid scenario. As we can note, overall, our socially inspired solutions present a better performance. When considering the delivery ratio (Figure 6.5a), for lower densities, we can observe that CC, Degree, CC-Degree, and ABSM deliver data messages to the same amount of vehicles. As the density increases, so does the delivery results for all solutions. However, for very high densities, the performance of ABSM and UV-CAST starts to deteriorate, while our proposals guarantee a 100% delivery ratio. In summary, this result shows that the considered social metrics lead to the same delivery capability.

Figure 6.5b shows the number of data messages transmitted. For lower densities, our proposals transmit more data messages than ABSM and UV-CAST. As shown in the previous result, given that CC, Degree, CC-Degree, and ABSM have the same delivery results for such lower densities, we can conclude that our proposals are not able to avoid redundant retransmissions when the network is sparse. Notice that the broadcast storm problem is not much of an issue in sparse networks. As the density increases, our solutions incur the lowest number of data messages transmitted. Among the three, Degree presents the best results, while CC gives the worst. Recall from the results shown in Figure 6.3 that at higher densities, the variability for the degree is higher than the one presented by the clustering coefficient. As already stated, the greater the variability, the greater the range of possible waiting delays, which leads to a better broadcast suppression approach. In a similar result, Figure 6.5c shows the number of collisions for all solutions. Essentially, the behavior is almost the same for the number of messages transmitted. Our approaches perform better at higher densities. It is worth noticing that at lower densities, among our solutions, Degree leads to the highest number of collisions, while CC leads to the lowest. This fact can also be explained by the variability results shown in Figure 6.3.

Figure 6.5d shows the delay for all solutions. As expected, for lower densities, the delay for all solutions is very high due to the store-carry-forward performed by all solutions; that is, vehicles need to store and carry messages around in order to deliver them. As the density increases, the delay

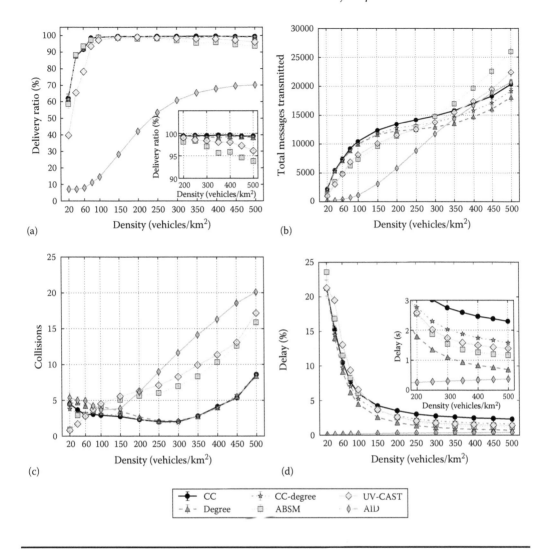

Figure 6.5 Simulation results for the Manhattan street scenarios.

for all solutions decreases. In particular, Degree has the lowest delay, while CC has the highest. According to the results shown in Figure 6.3, for higher densities, the clustering coefficient is also high. Therefore, the waiting delays chosen by vehicle will also be high, thus explaining the higher average delay. In the case of Degree, for higher densities, the node degree is also high. However, contrary to the clustering coefficient, nodes with a high degree have a lower waiting delay, which explains the average delay in delivering data messages to intended recipients.

The performance results for all solutions under the Cologne scenario are shown in Figure 6.6. As in the Manhattan grid scenario, our proposed solutions also have the best delivery results (see Figure 6.6a). In fact, CC, Degree, and CC-Degree are able to deliver about 15% more messages to intended recipients than ABSM and UV-CAST. This result shows that our proposals are more reliable than state-of-the-art solutions.

Figure 6.6b shows the total number of data messages transmitted. CC and CC-Degree transmit essentially the same amount as ABSM and UV-CAST, while Degree has a much higher overhead. Recall from Table 6.2 that the densities considered in the Cologne scenario are not very high. Moreover, as already shown, Degree induces a higher overhead under lower densities. Therefore,

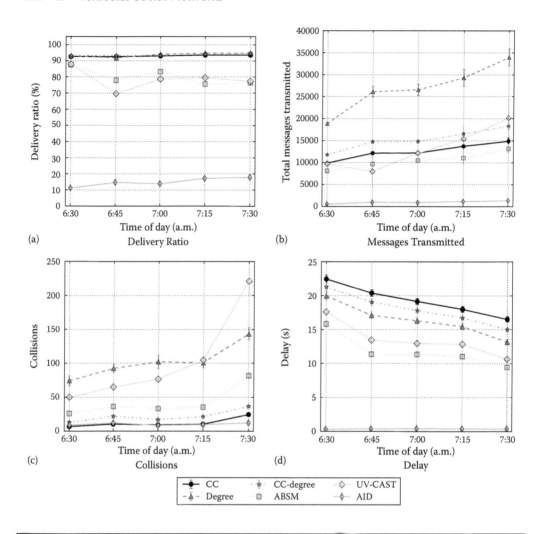

Figure 6.6 Simulation results for the Cologne scenario.

such a result was expected. The same behavior can be observed in the result of the number of collisions (see Figure 6.6c). Degree induces almost the same number of collisions as UV-CAST, while CC and CC-Degree generate the lowest number of collisions. As expected, CC behaves better at lower densities, which explains the better performance in the Cologne scenario. Figure 6.6d shows the delay result for the Cologne scenario. As can be observed, our proposed solutions have the highest delays. Such a result is intimately related to the delivery performance of our solutions when compared with the related solutions. That is, due to a higher delivery ratio (about 15%), and the fact that the Cologne scenario is sparse, vehicles need to store and carry the messages for a longer time, thus explaining the higher delay.

Finally, aiming to analyze the difference between the estimated cluster coefficient and the real cluster coefficient, we present the results for these metrics in Figure 6.7 under the Manhattan grid scenario. Recall that the estimated clustering coefficient is computed by considering the distance between vehicles and the estimated communication range. Therefore, signal attenuation caused by buildings has a direct impact on it. Conversely, the real clustering coefficient is calculated using the two-hop neighborhood knowledge of vehicles. As can be observed, overall, using the real clustering coefficient results in a better performance. However, the difference in the results of the estimated

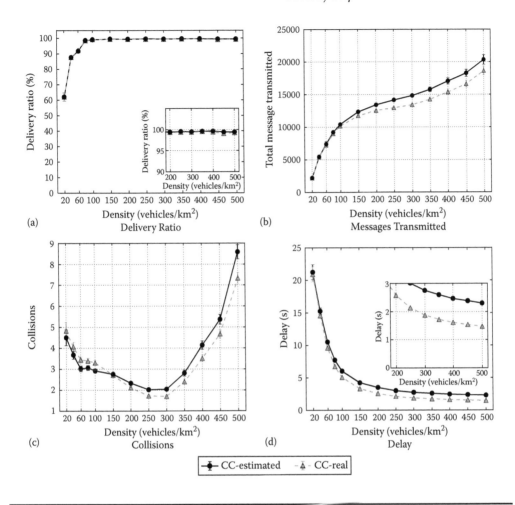

Figure 6.7 Comparison of performance evaluation between estimated and real cluster coefficient for the Manhattan scenario.

clustering coefficient is not significant, especially when we consider the extra cost to compute the real clustering coefficient.

6.5.5 Probabilistic-Based Results

Figure 6.8 presents the results for the Manhattan scenario. In Figure 6.8a, we can see the evolution of the delivery ratio when the density varies. It is possible to note that the CC-Degree and Topology solutions present a better performance, with values for the delivery ratio close to 100%. When compared with the other solutions, our solutions increase the delivery ratio in 5%. This happens due to the function to define the probability to rebroadcast the data message, which reduces the redundant transmissions. However, we can note a higher difference in the delivery ratio for the AID protocol, which defines its rebroadcast probability without considering the story-carry-forward mechanism. Figure 6.8b presents the number of messages transmitted. We note that in low-density regions, the CC-Degree and Topology solutions present a performance similar to that of the others. However, in higher densities, CC-Degree and Topology present less overhead than the other solutions, about 25% less. It is important to remark that the broadcast suppression mechanism reduces significantly the redundant transmissions.

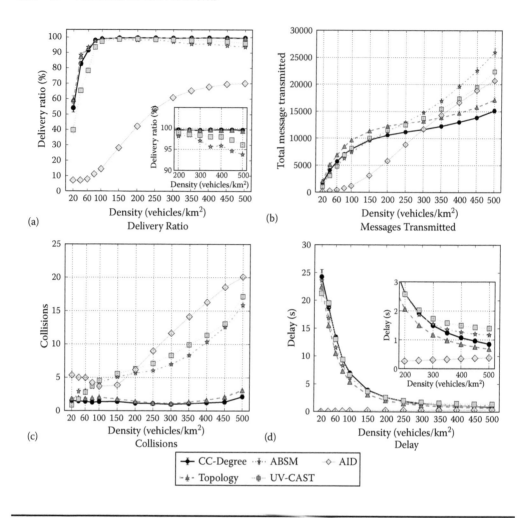

Figure 6.8 Simulation results for the Manhattan street scenarios.

When we look at the collisions, Figure 6.8c, we can note that CC-Degree and Topology present fewer collisions than the other solutions. Analyzing the density evolution, CC-Degree and Topology present a constant behavior in the curves, decreasing by 90% the total number of collisions. This occurs due to the fact that this is a scenario with fixed vehicle density in the whole area. It does not present network partitions, which favors the execution of the broadcast suppression algorithm, reducing the number of redundant transmissions. Regarding Figure 6.8d, we can observe the evolution of the delay during the dissemination. As we can see, when we compare the CC-Degree and Topology solutions with the delay-based solutions (ABSM and UV-CAST), the performance of our solutions is similar, presenting a reduction of 1 s in scenarios with higher density. Due to the fact that our solutions are probabilistic based, the vehicles do not add a delay in their transmissions, which contributes to this reduction. However, the AID rebroadcasts the data message directly, which reduces the delay but increases the collisions. Also, this behavior leads us to reach a low delivery ratio.

When we analyze the performance of the solutions in a sparse scenario (Figure 6.9, we note that our solutions (CC-Degree and Topology) guarantee a good delivery ratio, compared with the other

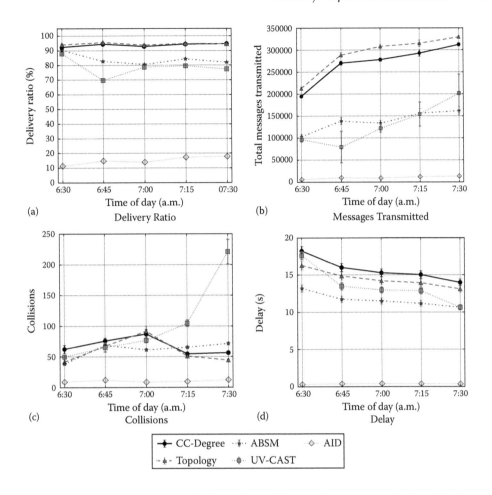

Figure 6.9 Simulation results for the Cologne scenario.

solutions (Figure 6.9a). We observe that the AID protocol presents the worst performance and the delay-based solutions present a delivery ratio of 80%. However, CC-Degree and Topology keep the delivery ratio close to 100%. Due to it being a sparse scenario, the store-carry-forward mechanism is necessary to guarantee a good delivery ratio. Figure 6.9b shows the results for the number of messages transmitted. As we can see, the proposed solutions present a great value compared with the other solutions. This is the expected result when we look to the delivery ratio. By being a sparse scenario, with a lot of partitions and low-density traffic, the solution needs to perform more transmissions to guarantee coverage of the whole ROI.

Observing the number of the collisions during the evaluated period (Figure 6.9c), we verify that the CC-Degree and Topology solutions present values close to those of the ABSM and UV-CAST solutions. Also, the AID protocol presents a smaller value of collisions due to the smaller number of transmissions. In this solution, few vehicles reach the probability to rebroadcast the data message. In Figure 6.9d, we note the results to the delay in dissemination. We note that CC-Degree and Topology present a delay slightly larger than that of the others. With the goal to guarantee the delivery in a sparse scenario, the tendency is that the delivery occurs with a greater delay. Once again, the AID solution presents a low delay due to the worst performance on delivery.

6.6 Conclusion

This chapter presented two new solutions to perform data dissemination in VANETs that take into account the social aspects of the network and the traffic evolution. Aiming to solve the broadcast storm and the intermittently connected network problems, we proposed delay-based and probabilistic-based solutions to perform data dissemination. The delay-based solutions use information about the number of neighbors (node degree) and how these neighbors are connected (clustering coefficient) as a criterion to define when and which vehicles rebroadcast data messages during the dissemination process. The probabilistic-based solutions, in addition to using the same metrics as the delay-based approaches, use the information of similarity of contacts, that is, the ratio of neighbors shared by two nodes (topological overlap), as a criterion to define the probability for each vehicle to rebroadcast the data messages.

We evaluated the performance of our approach in dense and sparse scenarios. In high-density scenarios, both solutions presented substantial performance gains, especially with regard to the delivery capability, delay, and overhead. Our solutions also presented a reduction in the last two metrics. However, in sparse scenarios our solutions did not present vast gains. Aiming to guarantee the delivery ratio in a scenario with partitions and variation in the density, our solutions transmit more messages, which contributes to increasing the delay and messages transmitted. We believe that this worsening in our solutions performance is the trade-off to guarantee good dissemination coverage.

As future work, we intend to improve our solutions by using other metrics better able to capture more social properties of vehicular networks, aiming to improve the selection criterion of rebroadcasting vehicles. Moreover, we intend to investigate mechanisms to detect and neutralize data injection attacks during data dissemination, aiming to increase the reliability of such a process.

6.7 Glossary

Data forwarder: Vehicles that receive data messages during the dissemination process and forward them to their neighbors.

Region Of Interest (ROI): An application predefined geographical region in which all vehicles inside it are intended recipients of the messages being disseminated.

Roadside Unit (RSU): Computing device placed on the roadside, which provides connectivity support for transiting vehicles.

Store-carry-forward: A data communication mechanism in which a vehicle receives a data message, stores it in a local buffer, and later forwards it to other vehicles as the holding vehicle moves around.

References

1. M. Bakhouya, J. Gaber, and P. Lorenz. An adaptive approach for information dissemination in vehicular ad hoc networks. *J. Netw. Comput. Appl.*, 34(6):1971–1978, 2011.
2. M. Behrisch, L. Bieker, J. Erdmann, and D. Krajzewicz. Sumo—simulation of urban mobility: An overview. In *International Conference on Advances in System Simulation (SIMUL '11)*, pp. 63–68, 2011. Barcelona, Spain.
3. A. Boukerche, H. A. B. F. Oliveira, E. F. Nakamura, and A. A. F. Loureiro. Vehicular ad hoc networks: A new challenge for localization-based systems. *Comp. Commun.*, 31(12):2838–2849, 2008.
4. F. Cunha, A. C. Viana, R. A. F. Mini, and A. A. F. Loureiro. Is it possible to find social properties in vehicular networks? In *IEEE Symposium on Computers and Communications (ISCC '14)*, pp. 1–6, 2014.

5. F. D. Cunha, G. G. Maia, A. C. Viana, R. A. Mini, L. A. Villas, and A. A. Loureiro. Socially inspired data dissemination for vehicular ad hoc networks. In *Proceedings of the 17th ACM International Conference on Modeling, Analysis and Simulation of Wireless and Mobile Systems, MSWiM '14*, pp. 81–85, New York: ACM, 2014.

6. E. M. Daly and M. Haahr. Social network analysis for routing in disconnected delay-tolerant MANETs. In *ACM International Symposium on Mobile Ad Hoc Networking and Computing (MobiHoc '07)*, pp. 32–40, 2007. Montreal, Canada.

7. P. O. S. Vaz de Melo, A. C. Viana, M. Fiore, K. Jaffres-Runser, F. Le Moul, A. A. F. Loureiro, L. Addepalli, and C. Guangshuo. RECAST: Telling apart social and random relationships in dynamic networks. *Perform. Eval.*, 87:19–36, 2015. Special Issue: Recent Advances in Modeling and Perform. Eval. in Wireless and Mobile Systems.

8. V. Drabkin, R. Friedman, G. Kliot, and M. Segal. On reliable dissemination in wireless ad hoc networks. *IEEE Trans. Dependable Secure Comput.*, 8(6):866–882, 2011.

9. M. Fiore and J. Härri. The networking shape of vehicular mobility. In *ACM International Symposium on Mobile Ad Hoc Networking and Computing (MobiHoc '08)*, pp. 261–272, 2008. Hong Kong, China.

10. T. Hossmann, Thrasyvoulos Spyropoulos, and F. Legendre. Know thy neighbor: Towards optimal mapping of contacts to social graphs for DTN routing. In *IEEE International Conference on Computer Communications (INFOCOM '10)*, pp. 1–9, 2010. San Diego, California, USA.

11. D. Katsaros, N. Dimokas, and L. Tassiulas. Social network analysis concepts in the design of wireless ad hoc network protocols. *IEEE Netw.*, 24(6):23–29, 2010.

12. X. Liu, Z. Li, W. Li, S. Lu, X. Wang, and D. Chen. Exploring social properties in vehicular ad hoc networks. In *ACM Asia-Pacific Symposium on Internetware (InternetWare '12)*, pp. 1–7. New York: ACM, 2012. Qingdao, China.

13. N. Loulloudes, G. Pallis, and M. D. Dikaiakos. The dynamics of vehicular networks in urban environments. In *CoRR*, abstract 1007.4106, 2010.

14. D. Naboulsi and M. Fiore. On the instantaneous topology of a large-scale urban vehicular network: The cologne case. In *ACM International Symposium on Mobile Ad Hoc Networking and Computing (MobiHoc '13)*, pp. 167–176, 2013. Bangalore, India.

15. F. J. Ros, P. M. Ruiz, and I. Stojmenovic. Acknowledgment-based broadcast protocol for reliable and efficient data dissemination in vehicular ad hoc networks. *IEEE Trans. Mobile Comput.*, 11(1):33–46, 2012.

16. R. S. Schwartz, R. R. R. Barbosa, N. Meratnia, G. Heijenk, and H. Scholten. A directional data dissemination protocol for vehicular environments. *Comput. Commun.*, 34(17):2057–2071, 2011.

17. C. Sommer, R. German, and F. Dressler. Bidirectionally coupled network and road traffic simulation for improved IVC analysis. *IEEE Trans. Mobile Comput.*, 10(1):3–15, 2011.

18. O. K. Tonguz, N. Wisitpongphan, and F. Bai. DV-CAST: A distributed vehicular broadcast protocol for vehicular ad hoc networks. *IEEE Wirel. Commun.*, 17(2):47–56, 2010.

19. S. Uppoor and M. Fiore. Insights on metropolitan-scale vehicular mobility from a networking perspective. In *ACM International Workshop on Hot Topics in Planet-Scale Measurement (HotPlanet '12)*, pp. 39–44, 2012. Low Wood Bay, Lake District, UK.

20. A. Varga and R. Hornig. An overview of the OMNeT++ simulation environment. In *International Conference on Simulation Tools and Techniques for Communications, Networks and Systems & Workshops (Simutools '08)*, pp. 1–10, 2008. Marseille, France.

21. A. M. Vegni and V. Loscrí. A survey on vehicular social networks. *IEEE Commun. Surv. Tutorials*, 17(4):2397–2419, 2015.

22. C. K. Verma, B. R. Tamma, B. S. Manoj, and R. Rao. A realistic small-world model for wireless mesh networks. *IEEE Commun. Lett.*, 15(4):455–457, 2011.

23. L. A. Villas, A. Boukerche, G. Maia, R. W. Pazzi, and A. A. F. Loureiro. DRIVE: An efficient and robust data dissemination protocol for highway and urban vehicular ad hoc networks. *Comput. Netw.*, 75:381–394, 2014.

24. W. Viriyasitavat, O. K. Tonguz, and F. Bai. UV-CAST: An urban vehicular broadcast protocol. *IEEE Commun. Mag.*, 49(11):116–124, 2011.

25. D. J. Watts and S. H. Strogatz. Collective dynamics of 'small-world' networks. *Nature*, 393(6684):440–442, 1998.

Chapter 7

Crowdsourcing Applications for Vehicular Social Networks

Elif Bozkaya and Berk Canberk

Department of Computer Engineering, Istanbul Technical University

Contents

7.1 Introduction

With the development of communication and computing technologies, the growing interest in the Internet of Things has resulted in a number of advanced technologies in wireless networks, such as smart cities, smart cars and smart homes. In this context, a variety of objects can connect and interact with each other in an ad hoc manner or over the Internet. One of the consequences of intelligent systems in vehicular networks is that vehicles can sense the environment and connect to other vehicles.

Cars are no longer simply vehicles. They are fast becoming high-speed Internet hot spots. For example, Audi and GM already have 4G long-term evolution (LTE) capability in some of their vehicles, but it is a technology that is becoming increasingly mainstream. AT&T and Audi announced that they planned to work together to have 4G LTE in every Audi 2016 model that is equipped with Audi Connect—a navigation system, Internet database, and in-car Wi-Fi hotspot. Ultimately, it means connected vehicles can communicate with each other and smart devices.

However, how can vehicles be connected and communicate with each other? More importantly, what are the advantages of all that connectivity in vehicular networks? To make a more intelligent world, a strong communications infrastructure is needed. AT&T's chief executive Glenn Lurie expressed the need for "better voice activation, better voice diagnostics—all the things you want so your hands stay on the wheel and your eyes stay on road."

One of the earliest technological advances was the ability to equip vehicles with on-board units (OBUs). Then, the U.S. Federal Communications Commission allocated 75 MHz in the 5.9 GHz frequency band for dedicated short-range communication (DSRC) in 1999. With the recent advances in wireless networks and the rapid growth of mobile communications and devices, vehicles are designed to increase situational awareness and improve traffic safety, efficiency, and comfort. Vehicles are equipped with advanced technologies in order to communicate with other nearby vehicles- so that vehicular networks have gained major importance.

Moreover, social networking has attracted all age groups by building virtual social communities. With the expansion of social networking during the last few decades, many social applications are being used through mobile devices. In particular, smartphones have become increasingly popular and are used to perform various daily routine activities. Therefore, social networking provides opportunities for users to connect or cooperate. Here, each user has a personal profile and relationship depending on social contents. Users have an opportunity to collect, process, and then share useful information with a group of people according to mutual interests and values. To this end, the usage of social networking has expanded in different fields of wireless networks due to the diverse capabilities of mobile devices and features of applications.

After that, recent advances in access technologies and increasing traffic density have triggered the emergence of many vehicular applications. Different types of applications aggregate the information from vehicles and make intelligent decisions. As a result, the applications and wireless technologies have resulted in better decision making and traffic management.

Vehicular social networks (VSNs) have emerged by characterizing the social aspects and features of the passengers and drivers in vehicular networks. VSNs provide mobile interaction between users on the road so that the efficiency of communication improves traffic safety. Moreover, if two vehicles have a common interest, in addition to communicating each other, they can make contact based on social metrics. More specifically, each user in the topology can be characterized based on friends, daily activities, routes, and so on and classified depending on social behavior. Vehicles can collect and elaborate useful data from users, depending on social groups.

However, the challenges address the interaction between vehicle and environment. How can social aspects coexist in vehicular networks? What is the importance of similar social characteristics in vehicular networks? What types of equipment and technology are required to interconnect the vehicles in vehicular applications? How can vehicles meet the quality of information required by the tasks in a dynamic social environment?

Here, the concept of crowdsourcing addresses many challenges and solves many complex tasks by harnessing the power of individuals. This phenomenon is especially impressive in VSNs. The observations of individuals and the capabilities of mobile devices are combined to obtain more effective solutions. The information is contributed by drivers and passengers through mobile devices and then used to achieve a complex task in vehicular crowdsourcing applications.

As a result, in this chapter, we study vehicular crowdsourcing applications based on the distinct characterizations of VSNs. Specifically, this chapter makes the following main contributions:

■ We briefly overview VSNs and then the challenges that affect the quality of vehicular crowdsourcing applications are investigated.

- To address the challenges in VSNs, we explain crowdsourcing paradigms in terms of mobile crowdsourcing networks (MCNs) and VSNs.
- We then investigate the existing vehicular crowdsourcing applications in VSNs.

7.2 Related Works

There is much research related to mobile crowdsourcing by utilizing the social context in VSNs. Crowdsourcing applications can be effectively used in VSNs for many purposes. However, many challenges significantly limit the performance of applications such as the task allocation of mobile crowdsensing and task execution in users' varying virtual communities and in the context of VSNs [4]. In [4], the authors propose a model to match multiple users with multiple crowdsensing tasks to execute the tasks efficiently. In [1], social-based vehicular applications are researched. In addition to current mobile applications in VSNs, new emerging technologies with next-generation vehicles and their main issues and challenges are also described.

On the other hand, several benefits of crowdsourcing are suggested. [13] focuses on a specific scenario related to community urban sensing applications. Mobile devices have sensors and wireless communication capabilities so that the level of accuracy can be evaluated by uploading data (e.g., traffic congestion, air pollution) to the Internet. To achieve this, a signal-processing approach is considered to solve the problem of evaluating the accuracy in a realistic scenario in a vehicular environment. [17] proposes a social network approach to sharing trustworthy information among vehicles. The authors create social connections between vehicles. Stranger vehicles are connected depending on the relationships in a VSN and then VSNs are constructed. The current research on direct and indirect trust in online social networks is discussed and research challenges in deploying trustworthy VSNs are addressed. [14] investigates crowdsourced data quality. In this work, the authors focus on estimating user motion trajectory information and propose a novel estimation method to differentiate normal users from abnormal users. The method enables more robust and accurate estimations of user motion trajectories and the mapping of fingerprints to physical locations for crowdsourcing-based mobile applications.

There have been many studies based on crowdsourcing in vehicular ad hoc networks (VANETs). [8] proposes a vehicular crowdsourced video social network for VANETs. When a user shares a video it can be accessed by other vehicles, and this includes the description of the video as well as views, comments, and likes. All these data collected from the video are stored in the social network cloud, and the model enables users to select appropriate videos on the road. The proposed architecture can be also implemented for safety-, efficiency- and comfort-related applications in VANETs. [11] presents a hybrid routing scheme for data dissemination using the location-based crowdsourcing of roadside units (RSUs) in VANETs. Here, the problem related to crowdsourcing is that the distribution of RSUs may change since Wi-Fi access points are deployed in a random manner on the road. Moreover, GPS does not give real-time updates of RSU locations. Therefore, this research proposes a probabilistic localization algorithm for VANETs using retrieval and crowdsourcing methods. [12] proposes a framework for context-aware and energy-driven route planning for fully electric vehicles via crowdsourcing. Due to the limitations of battery capacity in fully electric vehicles, energy efficiency is taken into account, and the impact of context awareness of real-time traffic on vehicle-to-vehicle (V2V) communication is evaluated via a crowdsourcing paradigm.

In addition, the impacts of social characteristics and common interests are researched in the literature to implement in vehicular networks. [15] investigates social interaction based on vehicular mobility. The authors characterize the mobility of vehicles according to social metrics such as distance, density, edge persistence and node degree. The described metrics aim to improve network performance and help drivers with vehicular applications. [16] proposes a social vehicular navigation system that enables drivers to share location-based traffic reports. Based on these reports, drivers obtain reliable information with which to select and plan their routes. The authors also present the system design implemented on an Android platform. [9] focuses on a specific scenario of VSNs and investigates the correlation between dynamic bandwidth allocation algorithms and social behaviors. The performance of the proposed algorithm and vehicular centrality are evaluated in terms of throughput and average queuing delay in VSNs. [10] proposes a probabilistic model by using trust rules in VSNs so that users can be administered and the interaction between users can be controlled. Here, social interactions in VSNs are grouped according to common characteristics and a dynamic trust management system is constructed. The model aims to minimize the impact of malicious behaviors in VSNs.

7.3 What Is Crowdsourcing?

Crowdsourcing is a combination of the terms *crowd* and *outsourcing*. In [6], it is defined as "the act of a company or institution taking a function once performed by employees and outsourcing it to an undefined (and generally large) network of people in the form of an open call. This can take the form of peer-production (when the job is performed collaboratively), but is also often undertaken by sole individuals."

Crowdsourcing has emerged to solve a wide range of tasks by harnessing the power of individuals. Crowdsourcing is being extensively used in both academia and industry with different aspects. How to cooperate, share and transmit data and users is addressed in different fields of science. For instance, there has been increasing interest in the field of computer science in terms of the crowdsourcing paradigm. The observations of individuals can enable companies and institutions to find a more effective solution to a complex task.

7.3.1 Mobile Crowdsourcing Network

Crowdsourcing is a distributed problem-solving model used by many organizations to solve a problem and then share the results. Here, mobile devices are in a suitable environment to complete the tasks and share with each other through the Internet. In such a case, MCN applies the principle of crowdsourcing to achieve tasks with human behaviors and mobile devices [5].

There have been notable changes in mobile and communication technologies with the evolution of the Internet. In particular, as mobile devices such as tablets and smartphones have become more widespread, most people tend to be online constantly. Figure 7.1 shows the percentage of smart devices and connections between 2014–2019. As seen in Figure 7.1, the total percentage will increase to 37% in 2016 and reach 54% by 2019 [7].

We see that mobile devices are getting smarter with higher computing resources. The huge percentage of usage of mobile devices results in increasing global mobile data traffic. This explosion of mobile data traffic and the proliferation of mobile devices affect the daily activities of people and their way of communicating. As a result, many mobile crowdsourcing applications have emerged with different purposes.

7.4 Vehicular Social Networks

For the last few years, vehicles have been equipped with different types of devices, such as GPS, radar, cameras, sensors and navigation systems to alert drivers to road conditions. As a consequence of advances in technology, vehicles are now more sophisticated.

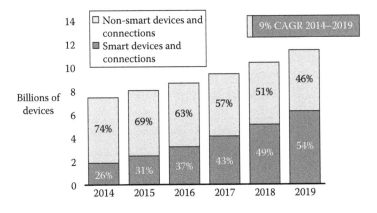

Figure 7.1 Global growth of smart mobile devices and connections [7].

While the basic task of vehicular networks is quite simple—enabling communication between vehicular nodes to improve traffic safety and efficiency— advanced technologies for vehicles are needed to achieve this. In addition to performing safety-related tasks, entertainment-related tasks are also added to vehicles.

Moreover, mobile social networks have gained major importance, whereby users in a community can share information and constantly stay in touch. Users rapidly acquire information and exchange data through the Internet. This has triggered the usage of mobile social networks in many fields of wireless networks.

The evolution of mobile social networks has led to advances in vehicular networks as well. VSNs are composed of vehicular networks and mobile social networks. Not only safety applications have improved traffic safety and efficiency and helped to avoid traffic accidents but entertainment applications have also advanced in the field of vehicular networks.

VSNs focus on social metrics between drivers, passengers, and groups in order to evaluate their performance. In this section, we briefly give some definition of the social metrics used in the literature as given in Table 1.1.

In VSNs, drivers can be classified into social groups according to common interests and relations among vehicles as shown in Figure 7.2. Drivers in a group can be interpreted in terms of sharing similar social interactions. The social groups include similar user profile information, including users' friends, routes, and activities. In general, relations among vehicles are based on similar travel routes, destination points and interests. Here, social interactions include messages about the traffic and road conditions. Drivers can use mobile devices to connect to the crowdsourcing application via the Internet. Then, the driving experience will be more comfortable, enjoyable, and safe by obtaining real-time information and exchanging data through the crowdsourcing paradigm.

For example, users moving along to the same destination point can exchange local topology information for traffic monitoring. Gathering information from the social group helps users in their daily activities. To this end, vehicles can minimize traveling time, avoid traffic congestion, or obtain real-time information on the road.

Figure 7.3 illustrates the architecture of aVSN. The system allows users with common interests to come together and form virtual communities. In addition, vehicles communicate with each other and RSUs and then obtain information with the help of OBUs for different types of applications. In addition, drivers and passengers collaborate through mobile devices.

Table 7.1 Social Metrics for Vehicular Social Networks

Social Metric	Definition	Evaluation Metrics
Degree centrality [9]	The popularity of a vehicle in terms of relaying	Queuing delay throughput
Distance [15]	The length of a path between a pair of vehicles	Number of hops
Density [15]	The amount of connections existent between vehicles	Density of edge
Edge persistence [15]	The persistence of an encounter between two vehicles	Mobility characteristic
Node degree [15]	The number of distinct encounters	Vehicle degree
Cluster coefficient [15]	A metric of how close are the neighbors of a vehicle in the graph	Cluster coefficient
Closeness centrality [15]	The centrality of the vehicle according to its distance to the other vehicles in the graph	Closeness centrality
Social interest profile [27]	The similarity of social interest between vehicular nodes	Probability that the connection is available

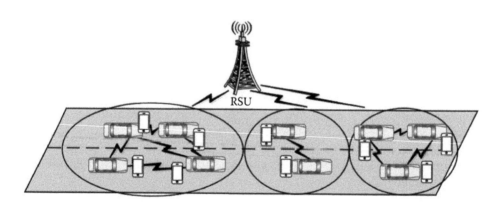

Figure 7.2 Vehicles are grouped based on common interests to obtain real-time information and exchange data through crowdsourcing applications.

7.4.1 Crowdsourcing for Vehicular Social Networks

Due to the sensing capabilities and properties of mobile devices, users have an opportunity to collect, monitor, and share data with neighbors anywhere at any time. Sharing personalized suggestions to users about routes, friends, activities, or events can help to schedule daily activities.

The drivers can use their mobile devices to access crowdsourcing applications via the Internet. For example, a location recommendation based on parking availability can guide drivers by preventing the ineffective use of time. However, this requires supplying growing demand and solving many complex problems. Here, these huge demands are met by the concept of crowdsourcing, which is a distributed problem-solving model and enables users to complete many tasks. The required information can be achieved via a crowdsourcing platform on the smartphone/tablet to

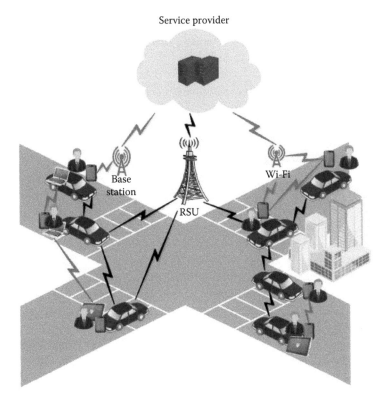

Figure 7.3 Vehicular social network architecture.

provide quick and real-time information. With these developments, many mobile crowdsourcing applications emerge in VSNs, such as traffic monitoring and tracking.

To support many vehicular applications, continuous transmission and network connectivity are needed. For example, in the United States, an average commuter drives approximately 26 km per day and spectrum occupancy characteristics vary while traveling on the road [2]. In such a case, many people travel for hours, especially when the traffic density is high and road conditions are poor. Thus, the usage of mobile crowdsourcing applications has gained major importance in vehicular networks.

In particular the increasing use of mobile devices and facilities in accessing the Internet enable social applications to extend to vehicular networks. Therefore, there is a growing interest in the usage of social applications in every field of wireless networks, including vehicular networks.

One of the main motivations of vehicular applications is to avoid traffic accidents and improve traffic safety and efficiency. Here, crowdsourcing is an essential platform to monitor road safety by utilizing the capabilities of mobile devices such as sensing, collecting, and sharing data. This can be achieved by establishing continuous communication and disseminating online information with robust network connectivity.

Mobile crowdsourcing enables users to solve specific problems in collaboration with each passenger and driver. Participants from the vehicular network environment efficiently contribute to improvement of network performance by exchanging information about road conditions or traffic congestion. Crowdsourcing gains major importance when the problem is a large-scale phenomenon and very complex.

In addition, knowledge of where vehicles will be in the near future enables traffic estimation and improves driving comfort. Here, crowdsourcing enables vehicle interaction according to social groups in VSNs. Therefore, vehicle location is one of the key parameters for many applications.

Moreover, crowdsourcing enables users to collect data from nearby vehicles based on social interactions and achieves more quality of information (QoI) for each user. For example, the connection of users in the social groups senses the environment, processes the sensory information by cloud computing, and makes intelligent responses for mobile applications.

7.4.2 Challenges of Vehicular Social Networks Related to Crowdsourcing Applications

The expansion of social networking creates both challenges and opportunities for vehicular networks. On one hand, vehicles can exchange messages related to safety, efficiency, and comfort to make better use of available network resources. On the other hand, due to high vehicular mobility, it is a challenge to meet the quality of service (QoS) of a task in crowdsourcing applications. Mobility models in vehicular networks are restricted by human behaviors and then formed by social characteristics. Therefore, the movement of vehicles is based on the observations of other drivers and the relations on the road. Here, the prediction of the future position of neighbor vehicles is a complex task. Obtaining real-time information is of major importance to evaluation of crowdsourcing applications.

In addition to the high mobility of vehicles, the limited transmission range of RSUs causes dramatic changes in the spatial and temporal behaviors of the network topology. These dynamic topological changes cause expressive degradations both in vehicle satisfaction and wireless communication quality in VSNs.

The features of VSNs are separated from traditional vehicular networks in many aspects. Here, one of the main challenges is to decide on the community structure. It means that deciding a community is to divide nodes into groups so that the nodes in each group are densely connected inside and sparsely connected outside [21]. Then, each vehicle is assigned into an appropriate group based on interests (similarity between nodes). Moreover, the social relationships are constructed between nodes. After that, vehicular nodes in a group communicate to accomplish various tasks collaboratively. All these steps can be achieved with intelligent algorithms with the help of a crowdsourcing paradigm. Some of the existing works are given in Table 7.2.

Today, mobile applications require continuous Internet connectivity, and then users exchange data through various wireless technologies such as Wi-Fi and Bluetooth. However, the increasing

Table 7.2   Proposed Solutions for Challenges in VSNs

Paper	Challenge	Proposed Solution	Evaluation Parameter
[4]	Matching between multiple users and crowdsourcing tasks	Application-oriented service collaboration model (ASCM)	Time delay battery consumption Network overhead
[4]	Dynamic contexts of users	Context-aware semantic service (CSS)	Effectiveness
[8]	Sharing video over VANETs	VeDi, for vehicular crowdsourced video social network over VANETs	Shakiness Blur
[26]	Ineffective data dissemination	A location-based crowdsourcing framework	Delivery ratio End-to-end delay

use of mobile devices causes the augmentation of mobile data traffic and data dissemination in a highly dynamic network topology, which can result in intermittent connectivity.

In addition to frequent changes in QoS requirements, the intrinsic restrictions of mobile devices such as limited screen size, keyboard size, processing time, and computing resources affect the user's comfort and quality requirements in crowdsourcing applications.

In many crowdsourcing applications, drivers share their locations automatically, which are stored in centralized servers. Then, this information is used in many different types of applications. In addition to obtaining location information, personal information such as profiles, friends, and messages is also collected by companies. This addresses the privacy challenge in crowdsourcing applications. To this end, several existing works focus on creating a trusted vehicular community [10,17].

To this end, there are many issues that should be considered. Mobility, human behavior, common interest, connectivity, privacy, and the limitations of mobile devices are some parameters to consider by evaluating network performance in VSNs.

7.5 Vehicular Crowdsourcing Applications

Vehicular networks are of paramount importance to drivers due to improving traffic safety and efficiency and provide a wide range of applications with different purposes. While safety is a major factor behind V2V and vehicle-to-infrastructure (V2I) communications, applications related to entertainment have also become popular in vehicular networks. Vehicular applications are generally divided into two main categories:

- ■ Safety applications: Safety applications are deployed to mitigate safety problems by decreasing traffic accidents and their severity in vehicular networks. When V2V communications potentially address about 4,409,000 crashes annually [3], the need for safety applications is clear.

  In safety applications, the delivery of a message in a timely manner and reliable information play a significant role. For example, in emergency situations, crashes are only prevented by the drivers' quick reactions and reliable, timely warning messages assist to improve personal response time and avoid traffic crashes. Hence, safety applications provide critical and important information collected by all vehicles.

- ■ Infotainment applications: With the recent advances in access technologies such as WiMAX, 3G, and Wi-Fi, infotainment applications have also emerged (e.g., video streaming, web browsing, voice-over IP). These applications especially have become more popular with the emerging applications with the aid of mobile Internet access.

With the consequences of the technological evolution, vehicles are becoming more sophisticated, with high computation, communication, and storage resources. Moreover, with the combination of mobile devices, vehicles can address many complex problem, through crowdsourcing. Crowdsourcing applications have greatly increased in recent years with the help of robust and continuous Internet connectivity in vehicular networks.

Many applications have used crowdsourcing as a method of acquiring, processing, sharing, and transmitting data with different purposes. Where did the traffic accidents occur? How fast is traffic moving on the road? What is the estimated arrival time depending on road conditions? All these questions are answered with real-time information and many applications help drivers in routing, time saving, reducing congestion problems, improving situation awareness and increasing the productivity of travel.

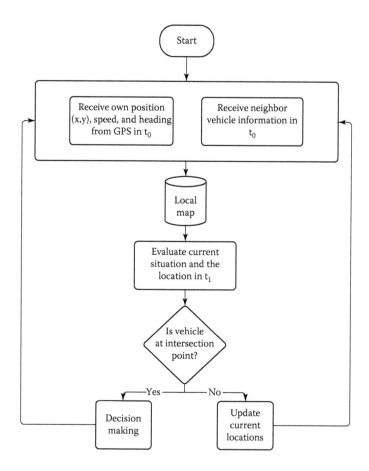

Figure 7.4   Decision making based on traffic and road conditions in vehicular crowdsourcing applications.

Vehicle locations are the most frequently exchanged information in VSNs. In particular, crowdsourcing applications are based on geographic information provided by GPS. Vehicles synchronize using GPS and decide the route in a timely manner. For vehicles traveling on the road, neighbor vehicle information is also required, as seen in Figure 7.4. Here, the local map involves vehicle information including position, speed, and heading and integrates with road and traffic conditions. A vehicle's current position is evaluated so that if it is at an intersection point, it can update the route for the destination point.

Traffic conditions are time-varying and affect the decision making of a vehicle. Collecting information from neighbor vehicles enables more effective and accurate knowledge of the road conditions. As seen in Figure 7.5, with the help of the vehicle's computation, communication, and storage resources, vehicular crowdsourcing applications can be used with different purposes, such as obtaining traffic information, road and weather conditions, and information dissemination. Depending on traffic and road conditions, drivers can schedule their trips and activities.

To this end, the remainder of this section addresses mobile crowdsourcing applications by giving specific examples.[*]

---

[*] Note that while all these applications require battery use on mobile devices, in-vehicle charge points are available that keep the devices charged continuously.

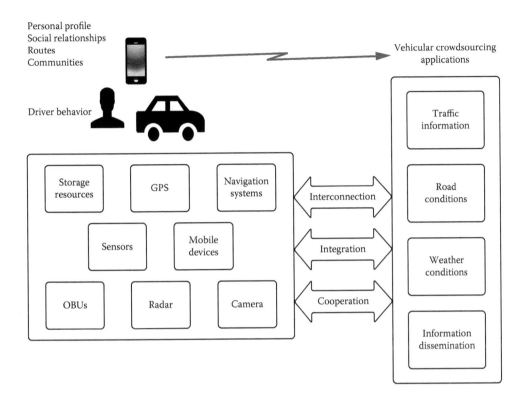

Figure 7.5 **Vehicular crowdsourcing applications can be used in order to acquire, process, share, and transmit data since the emerging technologies in automotive industries support real-time communications between vehicular nodes.**

Real-time data providers for traffic management are common for transportation agencies. INRIX [22] focuses on the idea that if you can analyze traffic information from the vehicles themselves, in addition to installing sensors on the road, you could understand road conditions everywhere. It analyzes the movement of people, vehicles, and commerce in cities, and provides real-time/predictive traffic information. The application and services help drivers with traffic maps, routing, and road alerts and also enable sharing with friends and family. Users can report accidents, hazards, and police activities. In addition, street parking and road conditions are available in the applications. TomTom [23] creates a clear and detailed picture of the traffic conditions and provides the most accurate and largest coverage in cities. The applications guide drivers to find the locations of people, and places and then route them in real time. TrafficCast [24] provides innovative technology and sophisticated data analysis for real-time and predictive traffic information that drives smarter on the road and enhances and enables high-performing location-based estimation and dynamic navigation services.

Here, social media such as Twitter and Facebook can be useful in unusual traffic conditions. For instance, users can share the information via social media and agencies can gather and disseminate information on the platform.

A popular example of a crowdsourcing service is Waze [25], a mobile application on smartphones. It helps drivers to connect to one another so that they can create local driving communities to improve the quality of daily driving. Users are able to report events that affect traffic conditions and remove the reported events after the event terminates in real time. Moreover, it navigates to the cheapest station on the route by enabling money saving. Similar to Waze, MapQuest [26] helps drivers to determine the best route between departure and arrival points according to the time of day, cost, weather, and situation. It gives an alternate route option when the traffic is heavy. The application can also be used to find a specific place, such as a gas station, post office, hotel, and so on. Drive and Share [18] is a social network application that allows drivers and passengers to exchange traffic and personal information on the road. One of the features of this application is that vehicles can participate in the social network automatically so that not only drivers and passengers contribute to the social platform, but also vehicles themselves can generate real-time information about the road conditions. Moreover, there are more applications based on real-time traffic information, such as NAVIGON, Scout, Sygic, Google Navigation, and Apple Navigation.

In addition, local public transport applications are also available on smartphones via crowdsourcing. For example, Moovit [27] is an intelligent local transit application to improve transit routes in cities worldwide. It guides users by giving nearby stations to destination points, compares transit options, and shows detailed directions all the way. It allows users to track their progress on the map and see the estimated arrival time, and sends alerts to users so that they can follow the trip in real time. Another example of a public transit information system is Tiramisu [28], which improves users transit experience and accessibility. It enables users to schedule trips, lists real-time arrival information, and indicates the fullness level of local public transportation. Users can see the upcoming stations and share experiences and suggestions.

7.6 Conclusion

In this chapter, we investigated the crowdsourcing application for VSNs. Crowdsourcing has the potential to address many problems in VSNs by gathering information from a group of people and delivering timely data to drivers. Here, connected vehicles and mobile devices solve the problem in a distributed manner. This chapter provides a taxonomy of the existing literature on the crowdsourcing paradigm for VSNs and then gives the challenges that affect the quality of vehicular crowdsourcing applications. Moreover, we explained the crowdsourcing paradigm in terms of MCNs and VSNs by giving specific examples of existing vehicular crowdsourcing applications.

References

1. A. M. Vegni and V. Loscri. A Survey on Vehicular Social Networks. *IEEE Communication Survey and Tutorials*, 17(4), (2015), 2397–2419.

2. R. Chen, R. Vuyyuru, O. Altintas and A. Wyglinski. *On Optimizing Vehicular Dynamic Spectrum Access Networks: Automation and Learning in Mobile Wireless Environments*, IEEE Vehicular Networking Conference (VNC), Amsterdam, The Netherlands, 2011, 39–46.

3. W.G. Najm, J. Koopmann and J. Brewer. *Frequency of target crashes for IntelliDrive Safety Systems*, U.S. Department of Transportation, National Highway Traffic Safety Administration, 2010. Washington, DC.

4. X. Hu and V.C.M. Leung. *Torwards context-aware mobile crowdsensing in vehicular social networks*, 15th IEEE/ACM International Symposium on Cluster, Cloud and Grid Computing, Shenzhen, China, 2015.

5. K. Yang, K. Zhang, J. Ren and X.S. Shen. Security and privacy in mobile crowdsourcing networks: Challenges and opportunities. *IEEE Communications Magazine on Security and Privacy in Emerging Networks*, 53(8), (2015), 75–81.

6. J. Howe. The rise of crowdsourcing. *Wired Magazine*, 14(6), (2006), 1–4.

7. Cisco Visual Networking Index. *Global mobile data traffic forecast update*, http://www.cisco.com/c/en/us/solutions/collateral/service-provider/visual-networking-index-vni/mobile-white-paper-c11-520862.html.

8. K. M. Alam, M. Saini, D. T. Ahmed and A. E. Saddik. *VeDi: A vehicular crowd-sourced video social network for VANETs*, 8th IEEE LCN Workshop On User Mobility and Vehicular Networks, Edmonton, Canada, 2014.

9. R. Fei, K. Yang and X. Cheng. *A cooperative social and vehicular network and its dynamic bandwidth allocation algorithms*, IEEE INFOCOM Workshop On Cognitive & Cooperative Networks, Shanghai, China, 2011.

10. N. Abbani, M. Jomaa, T. Tarhini, H. Artail and W. El-Hajj. *Managing social networks in vehicular networks using trust rules*, IEEE Symposium on Wireless Technology and Applications (ISWTA), Langkawi, Malaysia, 2011.

11. D. Wu, Y. Zhang, L. Bao and A. C. Regan. Location-based crowdsourcing for vehicular communication in hybrid networks. *IEEE Transactions on Intelligent Transportation Systems*, 14(2), (2013), 837–846.

12. Y. Wang, J. Jiang and T. Mu. Context-aware and energy-driven route optimization for fully electric vehicles via crowdsourcing. *IEEE Transactions on Intelligent Transportation Systems*, 14(3), (2013), 1331–1345.

13. M. Fiore, A. Nordio and C.F. Chiasserini. *Investigating the accuracy of mobile urban sensing*, 10th IEEE Annual Conference on Wireless On-demand Network Systems and Services (WONS), Banff, Alberta, 2013.

14. X. Zhang, C. Wu and Y. Liu. Robust trajectory estimation for crowdsourcing-based mobile applications. *IEEE Transactions on Parallel and Distributed Systems*, 25(7), (2014), 1876–1885.

15. F. D. Cunha, A. C. Vianna, R. A. F. Mini and A. A. F. Loureiro. *How effective is to look at a vehicular network under a social perception*, First IEEE International Workshop on Internet of Things Communications and Technologies (IoT'13), Lyon, France, 2013.

16. D. Kwak, D. Kim, R. Liu, L. Iftode and B. Nath. *Tweeting traffic image reports on the road*, 6th IEEE International Conference on Mobile Computing, Applications and Services (MobiCASE), Austin, TX, USA, 2014.

17. Q. Yang and H. Wang. Toward trustworthy vehicular social networks. *IEEE Communications Magazine on Security and Privacy in Emerging Networks*, 53(8), (2015), 42–47.

18. I. Lequerica, M. G. Longaron and P. M. Ruiz. Drive and share: Efficient provisioning of social networks in vehicular scenarios. *IEEE Communications Magazine on Converged Telecommunications Applications*, 48(11), (2010), 90–97.

19. D. Wu, Y. Zhang, J. Luo and R. Li. *Efficient data dissemination by crowdsensing in vehicular networks*, 22nd IEEE International Symposium of Quality of Service (IWQoS), Hong Kong, 2014.

20. T. H. Luan, R. Lu, X. (Sherman) Shen and F. Bai. Social on the road: enabling secure and efficient social networking on highways. *IEEE Wireless Communications*, 22(1), (2015), 44–51.

21. K. Xu, K. Zou, Y. Huang, X. Yu and X. Zhang. Mining community and inferring friendship in mobile social networks. *Neurocomputing*, 174(B), (2016), 605–616.

22. http://www.inrix.com/. Accessed 16 October 2016.

23. http://www.tomtom.com/. Accessed 16 October 2016.

24. http://www.trafficcast.com/. Accessed 16 October 2016.

25. https://www.waze.com/. Accessed 16 October 2016.

26. https://www.mapquest.com/. Accessed 16 October 2016.

27. https://www.moovitapp.com/. Accessed 16 October 2016.

28. http://www.tiramisutransit.com/t/livemap#reloaded/. Accessed 16 October 2016.

Chapter 8

Efficacy of Ridesharing as the Basis for a Dependable Public Transport System

Farzad Safaei

University of Wollongong, Australia

Contents

8.1 Introduction

Most modern cities face significant traffic congestion problems, with all the associated costs, such as environmental damage, psychological stress, and economic loss. Effective provision and use of public transport systems have long been considered the primary tools to reduce these externalities. However, the experience of the past several decades demonstrates that large-scale adoption of this mode of transport faces major hurdles. This is often despite increased efforts in modernizing public transport systems by introducing new and smarter technology, geographical information systems, and coordination among various modes of transport (e.g., [1,2]).

In Australia, for example, the number of private vehicles grew by an average of 4.2% during 1960–2010, much faster than the population growth (Figure 8.1), and the usage of public transport lagged behind that of personal cars by an increasing margin (Figure 8.2). This trend is likely to be repeated in many emerging economies.

Ironically, with the soaring number of cars on our roads, there is also a corresponding increase in the capacity to carry more passengers in the form of empty seats in these vehicles. Our congested cities, therefore, are faced with a paradoxical situation: *We are experiencing a shortage of reliable transport services while a massive oversupply of transportation capacity is choking our streets.*

For many drivers, the marginal cost of carrying an extra passenger would be small or zero, and they would be willing to offer a ride to others provided there were a system that could coordinate all the trips, guarantee safety, and eliminate potential awkward social interactions that may emerge. Many cities have been supporting carpooling by providing designated lanes or parking incentives, and there are many carpooling websites that attempt to help with the coordination of trips among the users. Nevertheless, the impact of traditional carpooling on traffic congestion has been insignificant, primarily because successful carpooling requires serendipity, negotiation, and coordination among people and a certain degree of trust.

More recently, however, the concept of carpooling has received a significant facelift and popularity boost through the use of advanced communications and smart phones. While traditional carpooling requires a prior agreement between the parties, it is now possible to find a suitable ride dynamically using the modern location-based services. The prime example is Uber [3], which is

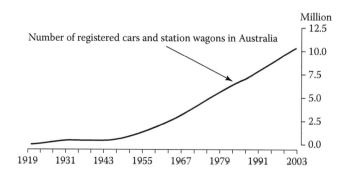

Figure 8.1   Growth in number of registered cars in Australia. (From Australian Bureau of Statistics [ABS], Motor vehicle census, 9309.0, Canberra: Australia.) [14]

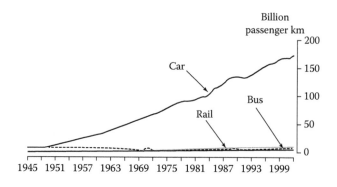

Figure 8.2   Usage of urban transport modes in Australia. (From Australian Bureau of Statistics [ABS], Bureau of transport economics, Working paper 38, Canberra: Australia.) [15]

now experiencing rapid growth and regulatory barrier fall internationally. The emergence of applications that utilize the so-called collaborative economy, such as Uber and Airbnb, is increasingly viewed in a favorable light compared with existing service models they seek to disrupt. Estimates by PricewaterhouseCoopers point to a massive increase in sharing economy revenues over the next 10 years, from $15 billion to around $335 billion [4]. In particular, innovations in sharing economy solutions may have significant benefits in the developing countries by easing infrastructure investment burden. Consequently, vehicular social networks are expected to be of significant benefit in the future [5].

The positive economic prospect has also resulted in a surge in research efforts to tackle various aspects of ridesharing and carpooling concepts. In the context of carpooling, Dimitrakopoulos et al. [6] developed a management functionality for dynamic ride matching by exploiting Bayesian networking concepts. Collotta et al. [7] tackled a key issue in carpooling, that is, a mechanism to increase trust, by prioritization of users based on their positive feedback. Significant research effort has been spent on efficient routing mechanisms for urban ridesharing. For example, He et al. [8] proposed mining GPS trajectories that can cope with traffic dynamics to yield shared routes for riders, while Cheikh et al. [9] focused on a multihop ride matching mechanism to improve the opportunity for finding a suitable ride. On the other hand, the interaction between ridesharing and existing public transport systems is also of interest. For example, Cangialosi et al. [10] introduced a generalized ridesharing system for connecting passengers to existing modes of public transport systems, and d'Orey and Ferreira [11] provided a case study for taxi sharing.

In this chapter, we focus on developing an analytical model for assessing the probability of obtaining a suitable ride, which to our knowledge has not been developed in the existing literature. We also consider a generalization of the concept of ridesharing consistent with the above-mentioned trend at large scale to tap into the underutilized transportation resources of the community. We now have all the necessary technological ingredients to create and properly *manage* what we refer to as the "transport commons." These ingredients include affordable devices that can run location-based services for passengers and vehicles, a reliable vehicular ad hoc network (VANET) augmented with a ubiquitous wireless backbone (3G, 4G, and so on), a generalization of the concept of social networking applied to spatially distributed users, and cloud-based services for calculation and coordination of routes. Our definition of the concept of transport commons is intentionally broad, so as to avoid being limited to a particular business model.

The key question addressed in this chapter is the effectiveness of transport commons to become a dependable complement to, or replacement of, the traditional public transport system. In particular, our aim is to derive an analytical model for the single most important metric of a public transport service from the perspective of consumers: *the probability of obtaining a suitable ride within a prescribed or desirable waiting time window*. The analytical and simulation results presented in this chapter demonstrate that the transport commons is indeed a strong candidate for a dependable public transport system.

The rest of this chapter is organized as follows. First, a brief description is provided of how a public transport service based on the transport commons concept may be implemented. Next, we develop an analytical model for the probability of obtaining a successful ride from the transport commons. Finally, we present simulation results and conclusions.

8.2 Description of the Transport Commons

The "commons" refers to any resource that is collectively owned or shared by the community. In the past, these were typically natural resources, such as the water supply of a village, fisheries, grazing grounds, or forests. Today, we also have artificial commons, such as works of art or software. Because our economic incentives are based on *private* ownership, humanity has always had difficulty with the management of commons, as exemplified by Hardin's "Tragedy of the Commons" [12]. Nevertheless, modern economic studies show that with suitable institutions for effective management, the commons can be a source of substantial benefit to society [13].

With current technology, it is now possible to create a transport commons management (TCM) entity that is accountable for safe operation of the system, able to enforce rules and moderate supply and demand, incentivize the right behavior, and weed out perverse incentives. TCM may be private or government owned.

People can use the transport commons as *passengers* or contribute to it as *drivers*. Of course, a given individual may be a passenger at one instance and a driver at some other time. Both passengers and drivers must register with the TCM, which may involve some procedures to ensure their good standing and suitability. Note that the ownership of vehicles remains with the drivers. The drivers donate any "trip" that they desire to the transport commons. We now describe the operation of the transport commons from the perspective of passengers and drivers.

Consider a passenger who intends to use the transport commons to get to work in the morning. The screen shots of two phases of the *TCM passenger app* running on the passenger mobile device are shown in Figure 8.3. The picture on the left presents the options that the TCM can offer the passenger. These options can include an integrated view of traditional public transport services and their timetable. If the passenger selects the transport commons option, the screen shot on the right may provide him or her with more detailed choices about the path of travel, the pickup point, the estimated arrival time at the destination, and the cost of the trip (if any). By choosing an offered path, the passenger is assigned to the chosen vehicle and the respective driver would be informed about the pickup location. The passenger can then make way to the pickup point and a handshake between the TCM apps on his or her mobile and the driver's device will confirm the right selection.

On the other hand, a driver who is registered with TCM can elect to donate any trip to the transport commons, for example, while going to work in the morning. A screen shot of the *TCM driver app* running on a suitable device in the vehicle is shown in Figure 8.4. During this trip, the driver app may use the device to establish a VANET and communicate with other transport

Figure 8.3   Example screen shots of the app used by the passengers.

Figure 8.4   Example screen shot of the app used by the drivers.

commons vehicles, as well as the TCM servers. The information gathered is used to calculate the expected delay incurred in choosing various paths, and also to suggest suitable passengers to the driver. In the example shown in Figure 8.4, two alternate paths are suggested by the TCM to the driver. The TCM may also provide an estimate of how many passengers are likely to be picked up for each path, as well as the associated estimated time of arrival. The driver is then free to choose among these alternatives. After the driver chooses his or her desired path, he or she is expected to follow the designated path to completion.

8.2.1 Economic and Social Considerations

By donating a trip to the transport commons, drivers will provide a community service and also enjoy lower congestion levels on the roads. They may also receive some financial benefits. For example, the cost of the trip may be tax deductible or the driver may receive a share of payments collected from the passengers. However, contributing to the transport commons, both as a driver and as a passenger, is beneficial to society at large, because it reduces congestion (a public bad) as well as the level of public investment needed to upgrade roads and other public transport systems. Consequently, it may even be economically desirable to reward both the drivers and passengers using a suitable mechanism.

The economics of the transport commons can be designed to provide incentives for drivers and passengers to adjust supply and demand. For example, a driver may be informed that there are many empty vehicles on the road today, and it is better to consider becoming a passenger. The incentives to become a passenger may be strengthened by some form of reward, as discussed above.

The drivers and passengers can also exercise some choice in the type of trip companion they will accept using social networking techniques. They may also be able to "rate" each other to help others with their choices.

8.2.2 System Architecture

Figure 8.5 shows the overview of the system architecture for this service. The TCM road status server collects various information, such as vehicle locations, their average speed, and congestion levels on the road. The information is used to obtain accurate estimates of road conditions and

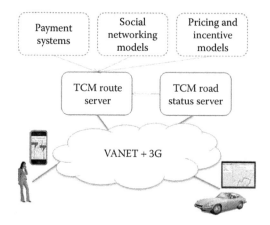

Figure 8.5   Possible system architecture for realization of transport commons.

is fed to the TCM route server. The TCM route server performs the appropriate route selection and assignment of passengers to vehicles. It will also provide recommendations on suitable travel paths to drivers. The TCM route server recommendations may also consider social and economic considerations in assigning passengers to drivers. For example, social networking options may be used to ensure that the social preferences of the users are met. Also, depending on the economic incentivization models used, the pricing information may impact these choices.

8.3   Analytical Model

A key metric that determines the suitability of using the transport commons is *service dependability*. A major contributor to service dependability is the *probability that passengers find suitable rides* to their destinations. We denote this probability as p_s and are interested in determining the important factors that influence this probability.

Let us consider a simplified model of a square city, as shown in Figure 8.6, where there is a road network in the form of a regular grid. There are $N \times N$ intersection points in this city. We use these intersection points as the vertices of the road graph, that is, places where new cars enter or leave the road network and passengers are picked up or dropped off. We assume that the arrival and departure of vehicles and passengers are uniformly distributed among the intersection points.

With respect to routes that vehicles adopt to reach their destination, we assume that traffic management at each intersection favors those vehicles that go straight over the turning vehicles. Consequently, to go from point b to e, the driver will use the two sides of the right-angled triangle, as opposed to the hypotenuse (Figure 8.6). This means that for each trip, there are two possible paths, shown as solid and dashed lines in the figure. Given the symmetry, the analysis of both paths will be similar.

Define a *unit of time* as the time that it takes for a vehicle to travel from one intersection point to the next. Using this unit, we divide the peak traffic period into a number of unit duration intervals. At a time interval t, denote the arrival rate of new vehicles whose drivers have donated their trip to the transport commons as $\lambda(t)$. We assume that the arrival of vehicles is uniformly distributed *in time* during the peak period, that is, $\lambda(t) = \lambda$ for all intervals t, which we believe is a reasonable assumption if the duration of the peak period is chosen to be relatively short.

We also need to consider the willingness of passengers to wait. Assume that passengers are only prepared to wait *one unit of time*, as defined above.

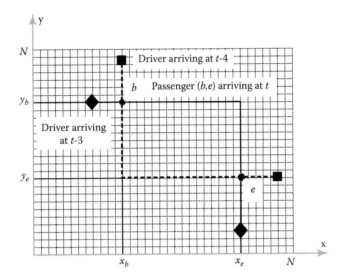

Figure 8.6 City with a regular grid road network. Two drivers are shown (black square and diamond) that can potentially pick up a passenger arriving at time *t* to intersection *b*. The passenger intends to go to intersection *e*, and alternative paths that drivers can take are shown as solid and dashed lines.

Consider now a passenger who arrives during the time interval t at the intersection b with coordinates (x_b, y_b) and intends to travel to point e with coordinates (x_e, y_e). Let s be a random variable representing the number of vehicles that have a *compatible* path with this passenger, that is, those vehicles that can pick up the passenger at b within the constraints of the waiting time and drop off the passenger at e *without altering their own path*. Clearly, s can only take nonnegative integer values. We can model the arrival of statistically independent vehicles as a Bernoulli trial, where success means the arrival of a vehicle having a path compatible with that of the passenger. Therefore, s will have a binomial distribution with parameters (n, p), where n is the number of trials (number of vehicles arriving) and p is the probability of success for each trial. We now intend to derive these parameters for our model. Let us first consider path 1 (solid line in Figure 8.6).

With respect to time interval t, the number of vehicles that enter the road network during this time interval is λ. The probability of success can be calculated by noting that only a subset of vehicles would be eligible to provide a ride to this passenger. This subset includes all the vehicles that satisfy the following conditions:

1. Their entering points are either (x_{b-1}, y_b) or (x_b, y_b). This is to ensure that the waiting time of the passenger is not more than one unit.

2. They have chosen path 1 for their journey.

3. Their destination points are anywhere between (x_e, y_e) and $(x_e, 0)$. This is to ensure that they can drop off the passenger without altering or extending their own path.

There are a total $2y_e$ such routes. The probability of each route can be calculated as follows. There are N^2 intersections in this city, and for a uniform distribution of arrivals, the probability of choosing a specific intersection is N^{-2}. Likewise, the probability of a specific intersection being selected as the departure location is N^{-2} for a uniform distribution of destinations. Therefore, the probability of each path (a specific arrival intersection *and* a specific destination intersection is N^{-4}.

However, assuming that each vehicle has an equal chance of choosing between the two available paths, only half of the vehicles will follow this path. Hence, if we define a random variable s_t as the number of suitable rides for this passenger from the vehicles arriving at time t, this random variable will have a binomial distribution with the following parameters:

$$S_t \sim B\left(\lambda, \frac{y_e}{N^4}\right) \tag{8.1}$$

Now let us consider those vehicles that are already on the road. There are λ vehicles that entered the system at $t-1$. Among these, only those with the following properties are able to pick up this passenger:

1. The vehicle entered the road system at (x_{b-2}, y_b) or (x_{b-1}, y_b). The arrival location is shifted back to ensure that this vehicle can pick up the passenger within the allowed waiting time, which is no more than one unit.

2. The vehicle chose path 1.

3. The vehicle had the same destination possibilities as before, that is, between (x_e, y_e) and $(x_e, 0)$.

This is a shifted pattern of arrivals by 1 compared with the vehicles arriving at t. Hence, we obtain the same binomial distribution for this case:

$$S_{t-1} \sim B\left(\lambda, \frac{y_e}{N^4}\right) \tag{8.2}$$

Continuing in this fashion for $t-2$, $t-3,\ldots$, we will eventually reach the city boundary (assuming that the duration of the peak period is long enough). There will be x_{b-1} of such random variables, each with an identical distribution as specified above.

We also have to consider vehicles arriving in the future. In this case, only the interval $t+1$ would be relevant due to the waiting time constraint. In addition,

1. The arrival point of such a vehicle must be x_b, y_b.

2. It must choose path 1.

3. It must have the same destination choices as above.

Therefore,

$$S_{t+1} \sim B\left(\lambda, \frac{y_e}{N^4}\right) \tag{8.3}$$

Hence, random variable s_1, representing the number of compatible rides using path 1, is the sum of these random variables. Given that all of these binomial distributions have the same probability of success and these random variables are independent, the distribution of s_1 is as follows:

$$s_1 \sim B\left(\lambda\left(x_b+\frac{1}{2}\right), \frac{y_e}{N^4}\right) \tag{8.4}$$

Similarly, the number of compatible rides using the second path (dashed line in Figure 8.6) can be derived as s_2. In this case, we need to replace x_b with $N-y_b$ and y_e with $N-x_e$, which results in the following:

$$s_2 \sim B\left(\lambda\left(N-y_b+\frac{1}{2}\right), \frac{N-x_e}{N^4}\right) \tag{8.5}$$

The desired random variable s is the sum of these two:

$$s = s_1 + s_2$$

Given that all x and y random variables have identical and independent uniform distributions with the mean of $N/2$, and the mean value of the binomial distribution for large n and small p is np, then the *expected value* of s for all passengers, denoted by μ, is

$$\mu = E[s] = \frac{\lambda(N+1)}{2N^3} \approx \frac{\lambda}{2N^2} \tag{8.6}$$

In the above expression, N^2 is a measure of the area of the city *in units of distance that corresponds to the acceptable waiting time of passengers*. Equation 8.6 shows that the mean number of compatible rides for a passenger is proportional to the ratio of the arrival rate of cars during the peak period to $2N^2$.

Using Poisson approximation of the binomial distribution, which would be reasonable for this problem (large n and small p), we obtain the distribution of a mean number of successful rides for passengers as

$$P(s) = e^{-\mu}\frac{\mu^s}{s!} \tag{8.7}$$

From the above, we can see that, on average, the probability of finding no suitable ride is

$$P(s=0) = e^{-\mu}$$

Hence, *the desired probability of finding a successful ride* is

$$p_s = 1 - e^{-\mu}, \text{ where } \mu \approx \frac{\lambda}{2N^2} \tag{8.8}$$

For example, consider a city roughly the size of Sydney (30 × 30 km) with 1 million vehicles arriving on roads during the peak period. Let the total duration of the peak period be 100 minutes, and 10% of drivers have donated their trips to the transport commons. Assume a grid size of 1×1 km^2 corresponding to a waiting time of around 1–2 minutes for typical city speeds. In this case, p_s would be approximately 43%. If 20% of drivers donate their trips, this is raised to almost 70%.

8.3.1 Assumptions and Discussion

The assumption of total independence and uniform distribution of destinations is probably a worst-case scenario. In a typical realistic situation, the traveling paths and destinations of people in a city are not completely independent of each other. One would expect to see some, perhaps strong, correlation between these in practice. For example, consider the other extreme, when everyone is traveling to a single destination, such as the center of a city. In this case, while the origins of their trips are still independent, everyone has the same destination. Using the same steps as before, we can derive the expected number of successful rides for this case as

$$\mu = \frac{\lambda(N+1)}{N^2} \approx \frac{\lambda}{N}$$

In this case, even if 1% of trips were donated to the transport commons, the probability of success would be around 97%, and with 3%, the probability would be almost 100%. The reverse direction would have the same behavior (i.e., all of the trips originate from one point, such as the center of city, and then the destinations are uniformly and independently distributed).

The real-world situation is likely to be somewhere between these two extremes; that is, there is some correlation between the origin and destinations of passengers. For example, in a typical modern city, there are points of confluence of people for work, shopping, or leisure, such as business parks, shopping centers, or sporting complexes. It is possible to change the analytical model to a finite number of origin or destination centers, as opposed to the N^2 that we assumed. The result is shown in Section 8.4.

In the above derivation, we assumed that the passengers were willing to wait at most a single time interval. We can use the same procedure to derive the probability of success if the waiting time were two or more intervals. We will present some results on the impact of waiting time in Section 8.4.

The above derivation did not impose any constraints on the number of passengers per vehicle. As the proportion of passengers to drivers increases, there will come a time when the probability of success is reduced due to this capacity limit. Nevertheless, assuming a capacity limit of two or three passengers per vehicle, the transport commons can provide a valuable service to a large number of residents with almost no infrastructure investment. In the example of Sydney presented before, 20% of drivers could carry in excess of 250,000 passengers during the peak period. This would reduce the queuing load on the road network significantly, with substantial reduction in congestion-induced waiting times. Note that the impact of reducing load on queueing delay is not linear. In particular, the rate of increase of delay will grow as the load gets closer to the capacity of the road network. Consequently, if the road network has been designed to cope with the peak period demand, reducing the peak traffic by around 25% would translate to a substantial reduction in congestion delay.

Note that the waiting time is obviously also affected by congestion on the road. In the model presented here, the cars travel from one intersection to the next in one unit of time deterministically. Extension of this model to statistical traveling time is left for future work.

8.4 Simulation Results

To verify the analysis and explore other situations, we have developed a simulation of 30×30 size grid city, as described in Section 8.3. In Figure 8.7, the probability of success is plotted against a range of arrival rates from $\lambda = 100$ to $\lambda = 1800$, which, based on Equation 8.1, corresponds to $\mu = [0.05, 1.0]$. As can be seen, the analytical model closely matches the simulation results but somewhat overestimates the probability of success. This discrepancy may be because the

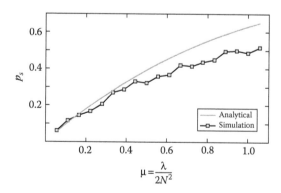

Figure 8.7 Probability of success as a function of μ, which is a measure of the number of cars donated to transport commons per second normalized based on the size of the city.

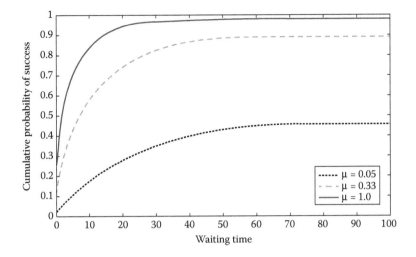

Figure 8.8   Probability of success as a function of waiting time of passengers.

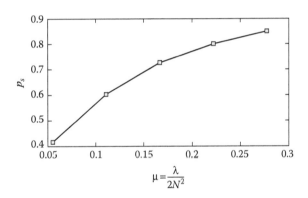

$$\mu = \frac{\lambda}{2N^2}$$

Figure 8.9   Probability of success where there are five centers of confluence of passengers.

analysis considers a very long peak period with many vehicles already on the road upon arrival of a passenger.

Figure 8.8 shows the cumulative distribution of success if the passengers are willing to wait. As can be seen for $\mu = 1$, almost all passengers will find a ride within 20 waiting periods.

Figure 8.9 shows a scenario where passengers and vehicles are equally likely to choose one of five possible destinations representing major business or commercial centers. The improvement of the probability of success in comparison with Figure 8.7 is substantial.

8.5 Conclusions

In this chapter, we introduced the concept of transport commons and described a possible system architecture for its implementation. We developed an analytical model to determine the probability of obtaining a ride from the transport commons by passengers. The results of the analysis, as well as simulation study, demonstrate that the transport commons is capable of providing a practical base for a dependable public transport service. In particular, even if only 10% of drivers donate their trip to the transport commons during the peak traffic period, the probability of obtaining a ride within a reasonable waiting period is quite high. The benefits of the system will also mean that contribution to the transport commons, both as a driver and also as a passenger, is huge. It may therefore be beneficial to the society if the transport commons is managed as other forms of commons, where this contribution is rewarded through a suitable mechanism.

References

1. B. Caulfield, M. O'Mahony. An examination of the public transport information requirements of users. *IEEE Transactions on Intelligent Transportation Systems*, 8(1):21–30, 2007.
2. W. Xiaolei, Y. Shunxi. Design and implementation of intelligent public transport system based on GIS. In Proceedings of the International Conference on Electric Information and Control Engineering, pp. 4868–4871, Wuhan, China, April 15–17, 2011.
3. http://www.uber.com (accessed December 20, 2015).
4. PricewaterhouseCoopers. http://www.pwc.co.uk/issues/megatrends/collisions/sharingeconomy/the-sharing-economy-sizing-the-revenue-opportunity.html (accessed December 20, 2015).
5. A. M. Vegni, V. Loscri. A survey on vehicular social networks. *IEEE Communications Surveys & Tutorials*, 17(4):2397–2419, 2015.
6. G. Dimitrakopoulos, P. Demestichas, V. Koutra. Intelligent management functionality for improving transportation efficiency by means of the car pooling concept. *IEEE Transactions on Intelligent Transportation Systems*, 13(2):424–436, 2012.
7. M. Collotta, G. Pau, V. M. Salerno, G. Scatà. A novel trust based algorithm for carpooling transportation systems. Presented at Proceedings of the 2nd IEEE Energycon Conference and Exhibition, Florence, Italy, September 9–12, 2012.
8. W. He, K. Hwang, D. Li. Intelligent carpool routing for urban ridesharing by mining GPS trajectories. *IEEE Transactions on Intelligent Transportation Systems*, 15(5):2286–2296, 2014.
9. S. B. Cheikh, S. Hammadi, C. Tahon. Based-agent distributed architecture to manage the dynamic multi-hop ridesharing system. Presented at Proceedings of the IEEE 13th International Symposium on Network Computing and Applications, Cambridge, MA, August 21–23, 2014.
10. E. Cangialosi, A. Di Febbraro, N. Sacco. On promoting modal shift via generalized ride-sharing. Presented at 2013 Proceedings of the International Conference on Connected Vehicles and Expo (ICCVE), Las Vegas, December 2–6, 2013.
11. P. M. d'Orey, M. Ferreira. Can ride-sharing become attractive? A case study of taxi-sharing employing a simulation modelling approach. *IET Intelligent Transport Systems*, 9(2):210–220, 2014.
12. G. Hardin. The tragedy of the commons. *Science*, 162:1243–1248, 1968.
13. E. Ostrom, R. Gardner, J. Walker. *Rules, Games and Common-Pool Resources*. Ann Arbor: University of Michigan Press, 2003.
14. Australian Bureau of Statistics (ABS). Motor vehicle census. 9309.0. Canberra: Australia.
15. Australian Bureau of Statistics (ABS). Bureau of transport economics. Working paper 38. Canberra: Australia.

Chapter 9

Security and Privacy in Vehicular Social Networks

Hongyu Jin, Mohammad Khodaei, and Panos Papadimitratos

Networked Systems Security Group, KTH Royal Institute of Technology, Stockholm, Sweden

Contents

9.1 Introduction

During the past decade, the trend has been to enable vehicle communication by equipping vehicles with on-board units (OBUs). Vehicular communication (VC) systems enhance intelligent transport systems (ITSs), enabling various applications based on vehicle-to-vehicle (V2V) and vehicle-to-infrastructure (V2I) communication. Vehicles can also access various service providers (SPs) through base stations (BSs), access points (APs), and road-side units (RSUs). Simply put, VC applications aim to provide awareness to avoid vehicle collisions and help drivers choose better routes based on traffic density [40]. This is achieved by active periodical beaconing of the vehicle's current status and sensed context information (e.g., obstacles or accidents), using OBUs preinstalled in the vehicles without any presumed relationship with neighbors (i.e., vehicles in the vicinity that receive messages while transmitting their own messages).

At the same time, interconnected vehicles facilitate message exchange beyond that for transportation safety and efficiency. "Socializing" with the drivers and passengers of nearby vehicles becomes possible. Unlike online social networks (OSNs) and most mobile social networks (MSNs), the users and devices (nodes) in vehicular social networks (VSNs) mostly interact when they are within communication range (i.e., physically close to each other, as determined by their trips). Due to the mobility of vehicles, they have ephemeral encounters and interactions. However, vehicle interactions could exploit such characteristics and even promote content dissemination in VSNs thanks to broad vehicle contacts during the trips. Moreover, vehicle connectivity to the Internet (leveraging BSs or APs) can enable interactions among VSNs, OSNs, and MSNs.

Traditional social networks leverage long-term user identities (i.e., an identity is created based on, e.g., an e-mail address or a username, and cannot be changed during its life span), and all user activities are carried out under these identities. The users also interact based on established relationships (e.g., with friends or group members), which are linked to their identities. However, these identities do not necessarily indicate the true identities of the users (i.e., user profiles can be fake). Thus, no strong authentication is required: the users are not kept accountable for their actions. On the contrary, in VC systems, there is consensus in academia and industry that vehicles should not expose long-term identities due to privacy concerns. Rather, short-term and unlinkable identities should be used to preserve user privacy. While privacy is important in VC systems, strong identification of drivers and vehicles is needed too, considering the high stakes in traffic systems (notably, driver and passenger safety). The messages in VC systems need to be properly protected, ensuring that the messages originate from legitimate users (nodes) in the system. Both requirements can be achieved leveraging pseudonymous authentication [8, 22, 26, 29, 31, 40, 41]. In fact, security and privacy in VC systems have been extensively studied, with significant efforts toward the deployment of secure VC systems, the basis for secure ITSs.

VSNs consider the VC network as the underlying networking facility, along with its location and context-specific services and features. While we embrace emerging VSN applications, it is important that VC system security and privacy are not compromised by the VSN functionality. VSNs could, and in fact should, be built upon the security infrastructures designed and deployed for VC systems and seek to address VSN-specific requirements based on extensions of those security infrastructures. Moreover, security solutions proposed for relevant functionality (e.g., location-based services [LBSs] and participatory sensing [PS]) could evolve and be integrated into VSN. This could largely promote the popularity and deployment of VSNs, rather than building the whole infrastructure from scratch; this is what we advocate in this chapter. We outline the VSN architecture and content dissemination in different architectures. We continue with the investigation of security and privacy requirements in the VC landscape. The awareness of those requirements is important in the context of VSNs, so that strong security and privacy protection for the overall system can be ensured while deploying VSN applications. Moreover, we survey the existing security and privacy solutions for emerging applications (potentially closely related VSN applications) and show that they could be integrated into VSNs. This eliminates the introduction of redundant components to the system. We close this chapter with a discussion of security and privacy challenges, and a brief conclusion.

9.2 Vehicular Social Networks

OSNs with rich features have been integrated into people's daily lives. They have satisfied user demand for socializing with friends or making new friends among people with common interests. Nowadays, OSNs are easily accessible from mobile devices (e.g., smartphones) and many of them exploit user mobility; thus, they are location-aware. Users can look for nearby users or tag posts with their current locations; this way, they can be discovered by other users through location-based searching.

OSNs maintain steady user relationships: users with common interests have direct or indirect relationships. Leveraging the Internet, user interactions are not time or space restricted. Although there exist decentralized social networks (e.g., Synereo*), the dominant OSNs (e.g., Facebook† and Twitter‡) are centralized, with their servers storing information about users or data generated and disseminated by them. Most OSNs follow a publish/subscribe model: users publish content to the central server and the server disseminates data to the users that subscribed to the content (e.g., followers or friends). Content dissemination in OSNs is not necessarily a real-time process: users can see their content at any time they wish to, as long as they have access to the Internet.

Social networks can also be decentralized. Decentralized social networks highlight user control over their own data. The data are stored locally and shared with other users they trust or closely relate to. Leveraging decentralization of social networks, user mobility could be exploited in MSNs to promote information sharing and region-specific interactions. Decentralized social networks emerged mainly due to privacy concerns in centralized OSNs [36, 49]. Centralized servers could breach user privacy simply because all user-related sensitive data are stored in them: data are exposed once the server is compromised, or even the server itself could be interested in the data. This coincides with privacy concerns in the context of VSNs, as will become clear in the discussion below.

Essentially, user socialization could appear in any network where communication (thus user interactions) is convenient. As described earlier, vehicles are expected to be communication-ready thanks to preinstalled OBUs. Thus, interconnected vehicles enable drivers and passengers to socialize with other nearby users, forming VSNs. VSNs inherit characteristics of traditional social networks, but they also have their own properties. In principle, VSNs are social networks built on top of vehicular ad hoc networks (VANETs), and they can be considered an extension of user-centric social networks. Applications in the VSNs could relate to a main purpose of VC systems, for example, transportation efficiency applications, while infotainment applications could be the main ones. We discuss the different characteristics of VSNs illustrated in Figure 9.1 in the rest of this section.

9.2.1 Networking Architecture

An OBU could integrate an IEEE 802.11p interface as well as cellular and Wi-Fi interfaces. Through the IEEE 802.11p interface, vehicles communicate with other vehicles (i.e., V2V) or

---

* www.synereo.com.
† www.facebook.com.
‡ www.twitter.com.

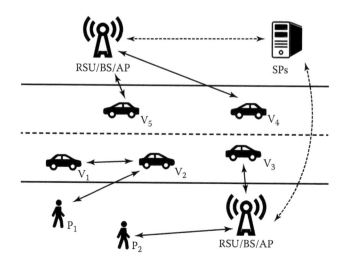

Figure 9.1 Illustration of VSNs. Vehicles (with OBUs, e.g., V_3, V_4, and V_5) and users (with smartphones, e.g., P_2) can access various SPs via RSUs, BSs, or APs. Vehicles (e.g., V_1 and V_2) or users (e.g., P_1) can interact with each other over an ad hoc network (e.g., share information obtained from SPs). (Icons made by Freepik from www.flaticon.com.)

with RSU (i.e., V2I). Cellular and Wi-Fi interfaces enable connection to the Internet via BSs and APs, and access to various SPs.

VSNs can be either centralized, decentralized, or in a hybrid form. Similar to OSNs, a centralized VSN involves a central server, and users can interact via that central server. Decentralized VSNs could leverage VANETs to form groups on the fly and enable communication among the group members. In addition, information obtained from the central servers could be shared with other nodes in an ad hoc manner.

9.2.2 Participation and Social Relations

The VSN participants are not limited to vehicles, but can also be passengers and pedestrians using smartphones or other portable devices. The VSNs primarily leverage VC systems while highlighting social connections between the participants. Smartphone users could bring OSNs and VSNs closer. In fact, data sharing among OSN and VSN applications could become more common. VSN users could share VSN or VC-specific data within other OSNs they join. Such interactions among OSNs and VSNs could promote the popularization of VSN applications. At the same time, smartphone-based ITSs [14, 23, 35, 51] have been proposed as an alternative approach before OBU-equipped vehicles become universal.

VSN applications could exploit different kinds of user relations. Interest-based applications could maintain long-term relations among users. VSN users could be friends in OSN applications and, at the same time, exchange VC-related and ITS-related information within the VSN context, for example, sharing traffic-related information. In this case, the OSN application only provides a way to establish relationships within VSN applications. VSN interactions could also be strictly specific to geographic regions, with these interactions being mostly short-term and not assuming any prior relationships. For example, passengers with the same destination (in public transportation or in private vehicles) can share point of interest (POI) information around the destination.

9.2.3 Applications

Any transportation-related information exchange could be facilitated by VSN applications. Users can obtain traffic information from central providers and share this information with nearby users, or they can sense surrounding traffic conditions and construct a broader (regional or, in a sense, global) view based on contributed sensing data from multiple nearby users. This can be seen as an extension of VC applications. LBS is another type of important application in VSN. Traditional LBSs (e.g., querying POIs from LBS servers) could still exist in VSNs. However, as socialization is highlighted in VSNs, users could largely exploit their mobility and even generate customized location-dependent information (e.g., travel guide for a certain place) based on their interests, which can be later shared with other users (with similar interests). This is more dynamic and more user-centric than information that could be obtained from traditional LBSs. In general, users can in fact share any information they want to, within VSNs, as long as the senders and the receivers are allowed to do so (e.g., not copyrighted music).

9.3 Security and Privacy Considerations in VSNs

Security and privacy are key factors for designing and deploying a large-scale trustworthy VSN. As described earlier, VSNs are built on top of VC systems and VSN applications should not deteriorate the achieved VC system security and privacy. Security and privacy requirements could vary depending on VSN applications. For example, for a safety application (e.g., hazard warning), integrity, nonrepudiation, and accountability are of paramount importance (unlike confidentiality), whereas for a traffic efficiency application, not only the integrity but also the verifiability of the content is crucial to prevent users from being misled. On the other hand, for an infotainment application, the availability of the service is important. Next, we list and explain the basic security and privacy requirements for VSNs based on those for VC systems [41]; in Section 9.3.2, we further explain security and privacy concerns in terms of possible adversaries in VSNs.

9.3.1 Basic Security and Privacy Requirements

Next, we list the security and privacy requirements for VC systems and VSN applications.

Authentication and integrity: A node should authenticate the source of a message so that only the information from trusted, that is, legitimate, sources is accepted. Moreover, messages should not be tampered with: unauthorized entities should not be able to alter the content of the messages.

Confidentiality: Information exchanged by the users should be confidential, that is, accessible only by authorized recipients, for example, vehicles in a platoon or vehicles from the same manufacturer. Information could, nonetheless, be simply broadcasted, and thus there is no need to be confidential (e.g., traffic conditions disseminated to a specific region).

Accountability and nonrepudiation: Entities in the system, including vehicles (i.e., OBUs), mobile devices, and infrastructure entities, should be accountable for the actions within the system, and they should not be able to deny the actions they have performed in the system.

Unlinkability and anonymity: User identities should not be exposed; that is, users should be anonymous and their (authenticated) messages should not be linkable. However, for practicality and efficiency, we inherit conditional anonymity (i.e., pseudonymity) from the VC domain: user messages can be linkable over a system-defined period τ, and users can remain pseudonymous as long as they do not misbehave. Moreover, users could opt to gain and accumulate reputation or credits for their contribution to the system, while using (long-lived) pseudonyms as their legitimate identities in the system.

Access control: Only legitimate entities, registered within the system, should be able to operate and contribute to the system. The system should prevent any illegitimate entity from participating in system operations, for example, content delivery or crowdsourcing. In VSNs, user interactions could also be restricted by relationships: unlike message broadcast and sender authentication in VC systems, user interactions could be allowed strictly based on relationships (e.g., among friends).

Availability: The system should remain operational even in the case of a fault. In particular, the underlying VC network architecture (i.e., user safety and traffic efficiency) should not be affected by the VSN traffic.

9.3.2 Adversarial Model

9.3.2.1 Honest but Curious Entities

Recent experience from mobile applications (e.g., LBSs) [25] shows that SPs are aggressively collecting user information in order to profile users. For example, an LBS server could collect user queries (including user locations and interests) to offer customized services or push advertisements to the users. This raises user concerns regarding their privacy. More generally, every entity within the system, for example, passive observers, SPs, and security infrastructure entities, could infer information in order to infringe user privacy. A number of works try to solve this problem by transferring the trust to an additional trusted third party (TTP) [18, 37] introduced in the system: a proxy between the users and the honest but curious server, so that all user requests are anonymized by the proxy before forwarding to the servers. However, the same concern applies to any such entity: if the same information is available to the servers and the introduced TTPs, then there is no difference between what the servers and those entities could do (i.e., the information they can infer).

This is why we need to extend our adversarial model from *fully trustworthy* to *honest but curious* servers. Honest but curious entities never deviate from system security policies or protocols, but they are tempted to infer and exploit user-sensitive information, for example, profile users and push advertisements to users based on their interests.

9.3.2.2 Malicious Participants

Due to the dynamic nature (intensified in a decentralized architecture) of VSNs, registered vehicles and users (legitimate insiders) are able to disseminate faulty information to affect a process, for example, measurement of traffic conditions. Internal adversaries may also tamper with the content they obtain from a provider before they share it with other users. The openness of VSNs leads to additional vulnerabilities compared with those in traditional social networks. Polluted data reported from faulty insiders should be filtered out, and malicious users should be evicted from the system. This requires accountability of user actions. The situation could be worse if a malicious user were able to equip herself with multiple seemingly valid identities and affect the system. For example, an adversary could clone an identity (which she should not own) to mislead other users by disseminating aggressively false information. Such Sybil-based misbehavior [12] can be the basis for various kinds of attacks, for example, injecting bogus messages to control the outcome of a specific protocol, or disseminating spam to other users.

On the contrary, external adversaries have relatively limited capabilities. They can always try to harm user privacy by eavesdropping on wireless communications, or they could simply launch jamming and distributed denial-of-service DDoS attacks on a specific target (e.g., server) or area to undermine the system availability.

9.3.2.3 Selfish Participants

Crowdsourcing-based mobile applications [15, 16, 34, 46] have been widely used for enhancing transportation efficiency and safety. These applications rely on user participation and contribution to measure specific phenomena (e.g., traffic status). However, such applications would not work without the active participation of users. In a VSN, selfish users could try to achieve higher rewards by sacrificing the minimum resources. These misbehaving internal adversaries would rely on resources of other nodes to benefit from the VSN services without actually contributing to the tasks [28]. The success of such applications depends on the participation of (the majority of) users and their collaboration to achieve the desired goals. Unless mechanisms that motivate or incentivize user participation are in place, selfish users would not be willing to consume resources. The system should be able to identify selfish users or free riders, and then degrade and limit their access to (the services in) the system, or even evict them altogether.

9.4 Existing Security and Privacy Solutions

A lot of research on security and privacy has been carried out in relevant areas, VC systems, MSNs, and crowdsourcing. Security and privacy solutions in those areas could evolve and be customized to address similar problems that exist in the VSNs. In this section, we discuss such solutions that could facilitate secure and privacy-preserving VSNs.

9.4.1 Decentralization

Decentralization could be motivated by various reasons, including privacy. Honest but curious servers in OSNs or LBSs tend to collect sensitive user information and infer more from the collected data [20, 27, 45]. Location privacy is a main concern in VSNs: interactions among VSN entities are location dependent. Information obtained from the servers is customized based on the vehicle location (e.g., LBSs). Geographical information could be used to track the users and even infer the interests of a specific user from the information being requested. *k-anonymity* [18, 19] has been widely used for protecting location information in both centralized and decentralized settings. Anonymizer-based approaches leverage an anonymizer introduced between the users and the servers [18, 37]. However, such an anonymizer could also be a threat for user privacy, also being honest but curious [27, 45]. Decentralized approaches have been proposed to eliminate such concerns. Users could leverage peers around them to form a region that involves at least $k - 1$ other users and use this obfuscated region instead of an accurate location [19, 24]. Such approaches trade off user overhead, to search for nearby peers for privacy protection. In fact, the peer selection strategy determines the efficiency and effectiveness of such schemes [19, 24], which is harder, especially when the node (peer) mobility is not predictable.

Such collaboration in VSNs could be made easier by forming groups in VANET leveraging similar vehicle mobilities. Nearby vehicles could form groups and maintain them as long as the vehicles are within each other's communication range [43]. Each group has a leader that acts as a temporary anonymizer for the group. The group leadership is rotated over time to share the burden among the group members and limit the information the group leader could learn. Such temporary centralization leverages the VANET characteristics and decreases the overhead to search for the most suitable peers. However, these approaches would not help if the honest but curious server is only interested in symbolic locations (e.g., church, shopping mall, and railway station) of the users: in that case, all k members are very likely to be at the same symbolic location.

Content sharing can further protect user privacy, because they would query the content provider less frequently, thus reducing their exposure. For example, information sharing in LBSs [45] helps users protect their privacy in a collaborative way: LBS-obtained information is shared with their neighbors (in terms of peer query and response) and the LBS server is queried if no satisfactory peer response is obtained. However, this allows internal attackers to provide faulty information to benign users; receivers do not have a clue whether the information is valid, as long as they do not query the LBS server directly. Thus, user authentication is needed to eliminate illegitimate users from the network [27].

9.4.2 Pseudonymous Authentication

To thwart vulnerabilities due to the openness and decentralization, transmissions in VSNs should be verifiable in terms of trust, especially for the safety-related applications. In most of the VSN applications, users may be mostly strangers to their peers without prior social interaction. Vehicles have limited time to share information with each other due to mobility. This implies that it is hard for users to accumulate a reputation for trust establishment. Public-key infrastructure (PKI)-based solutions could be used to ensure the authenticity and integrity of the transmitted messages, leveraging a TTP (i.e., a certification authority [CA]) to establish trust among vehicles. However, with traditional certificate-based authentication, one can easily trace the messages transmitted by a specific vehicle based on its identity (in the certificate), and thus profile its behavior, especially considering the openness of wireless networks. Message encryption would help so that only the targeted recipients could decrypt the messages. However, as described earlier, vehicles have ephemeral encounters; thus, it would be unrealistic to negotiate (multiple pairs of) security associations within short periods with (multiple) recipients and encrypt all transmissions. Moreover, it is hard to decide in advance the interested recipients in case transmissions are region based or targeted; that is, the messages should be authenticated and broadcasted to all neighboring nodes. Thus, the approaches relying on long-term identity cannot be used, as all user actions would be linkable. This motivated solutions leveraging anonymous credentials to satisfy both security and privacy requirements in the VC domain.

There are two categories of vehicular identity and credential management schemes proposed for VC systems: public key-based and group signature-based schemes. The public key-based schemes [1, 2, 22, 29, 44, 48] propose to equip vehicles with a set of short-term pseudonymous credentials (termed pseudonyms). A pseudonym is a public key authenticated by the pseudonymous certification authority (PCA). The pseudonyms are essentially unlinkable; that is, one cannot link two pseudonyms, as they do not include any information that could be linkable. Each user signs outgoing (time- and geo-stamped) messages using the private key corresponding to the current valid pseudonym. The pseudonym (and possibly the chain of trust) is attached to the messages in order to facilitate verification by the recipient. Depending on their spatial, temporal, or interest scope, receivers verify the attached pseudonyms first and then validate the signature using the public key of the attached pseudonym.* The vehicle switches to another pseudonym regularly, for example, based on pseudonym validity periods, and signs messages under the new pseudonym. Using pseudonyms, one can achieve integrity, nonrepudiation, accountability, and conditional anonymity. Pseudonyms

---

* We assume that the sender and receiver trust the pseudonym issuer, the PCA.

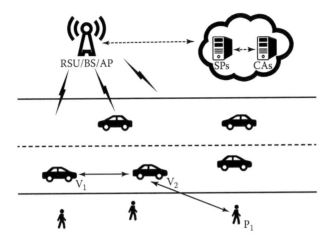

Figure 9.2 Vehicles and users can obtain pseudonyms from the CAs. The communication in the VSNs is protected with pseudonymous authentication, including peer-to-peer (P2P) communication (e.g., V_1-V_2 and V_2-P_1) in the ad hoc network and vehicle user–SP communication. (Icons made by Freepik from www.flaticon.com.)

can be integrated with different services (and their SPs) to provide secure and privacy-preserving VSNs. For example, in [27], a secure and privacy-enhancing LBS is proposed leveraging information sharing and pseudonymous authentication. Users authenticate their queries and responses under the pseudonyms. This way, illegitimate users are prevented from providing false information, while internal adversaries are kept accountable for their actions. Figure 9.2 illustrates a secure and privacy-preserving VSN architecture leveraging pseudonymous authentication.

On the other hand, anonymous authentication schemes [7, 32, 33, 48] were proposed for VC systems. These approaches leverage group signatures [3, 4, 9]: a receiver can verify that a legitimate group member signed a message without knowing who the signer is. In case of misbehavior, the group manager is able to *open* the signature, thus disclosing the signer's identity and revoking it.

While group signatures can be used to protect transmitted messages, such schemes incur high computational overhead for signature generation and verification [38]. The integration of group signatures and pseudonyms could decrease the computational overhead for resource-constrained devices. A hybrid approach, with the vehicles generating public or private key pairs and signing the public keys with their own group signing keys, was proposed in [6]. A pseudonym needs to be verified only once, the first time it is received, and cached locally during its lifetime. For subsequent messages signed under the cached pseudonyms, only the signatures on the messages need to be verified, using the traditional public key, which is much cheaper than group signature verifications.

9.4.3 Sybil Resilience

It is possible that a compromised vehicle equips itself with multiple simultaneously valid pseudonyms (e.g., by requesting pseudonyms for the same period multiple times). This sets the ground for Sybil-based misbehavior [12]. Sybil resilience in VC systems remains an open challenge in the absence of consensus because the standardization bodies [13, 26] and harmonization efforts do not have conclusive views on that front. For example, Car2Car Communication Consortium (C2C-CC) [8] proposes issuing pseudonyms with overlapping lifetimes in order to keep the safety applications operational at any given point in time [30], while [29] and [40] (works in the context of SeVeCom [31] and the PRESERVE EU project [47], respectively) propose issuing pseudonyms with nonoverlapping lifetimes in order to eliminate the possibility of equipping a vehicle with multiple simultaneously valid short-term identities.

Each vehicle can be equipped with a tamper-proof hardware security module (HSM), to prevent adversaries from manipulating pseudonym acquisition and usage [22,39]. A more straightforward solution would be to have the pseudonym provider simply issue pseudonyms with nonoverlapping lifetimes and keep a log of pseudonym issuance to reject spurious requests. However, this would not work if there exist multiple pseudonym providers (an adversarial vehicle could request pseudonyms from different pseudonym providers). Moreover, pseudonym providers could not share records (otherwise, they could track vehicles) [20,29]. The state-of-the-art proposal is to separate duties with one (or a few) identity provider and multiple pseudonym providers [20,29,30]. An anonymized ticket is issued by the identity provider to enable the vehicle to obtain pseudonyms from a pseudonym provider. Each ticket is bound to a specific pseudonym provider, without the identity provider learning the pseudonym provider. Each ticket can be used only once, while not revealing the location of the vehicle to the identity provider. Moreover, the identity provider only learns the period a vehicle has requested pseudonyms for, so no ticket will be issued again for the same period, which further prevents a vehicle from equipping with multiple simultaneously valid pseudonyms.

9.4.4 Data Verification

Entity authentication would help in eliminating illegitimate users and enhance the trustworthiness of shared content. Leveraging postmisbehavior approaches (e.g., pseudonym resolution [29]), entities that provided faulty information could be "punished" (e.g., revoked from the system). However, such a reaction presumes that misbehavior is already detected, which is not trivial.

The authors of [11, 21, 42] propose internal attacker detection approaches for sensing data aggregation based on redundancy of data received from multiple sensing entities, assuming the majority of internal nodes are honest. In all works, the authors leverage entity authentication. In [11], each vehicle aggregates received data corresponding to the same event and merges them with its own sensing data, and then forwards the aggregate to neighboring vehicles. Each received message contains a path list, and the redundancy is determined based on the nodes included in the path lists so that malicious nodes could not increase the redundancy of false information by affecting the aggregated data from multiple paths. In [21], data are sent and aggregated by a server. The aggregated data then can be queried by the users. The server is trained with initial submissions and detects outliers purely based on the sensing data (e.g., temperature measurement) submitted from users' mobile devices. Raya et al. [42] propose data-centric trust establishment in ad hoc networks. Trustworthiness is decided based on messages instead of message generators. Bayesian inference and the Dempster–Shafer theory of evidence are used for evaluation and decision making with various metrics (e.g., proximity and time).

9.4.5 Incentives

Collaboration is the basis of security and privacy solutions in various domains (e.g., privacy-enhancing LBS [45] and PS networks [5,17,20]). LBS-obtained information is shared with other peers looking for the same information, so that they do not expose their location (and activities) to the possibly honest but curious LBS servers [45]. Crowdsourcing applications leverage user contributions in the system to infer context-dependent data (e.g., temperature and humidity). In principle, user experience would be improved with increased user participation. However, selfish users could choose to benefit from the system without contributing to the system (e.g., requesting information from neighbors while not sharing information with others).

Incentive schemes can be used to ensure user contribution, in which users are awarded virtual credits for their contribution. Zhou et al. [52] propose an incentive mechanism for data forwarding in a vehicular delay tolerant network (DTN), in which a central server stores the credits of different users. After each successful transmission, the credit is charged from the source node and distributed to intermediate nodes that relayed the packets. If a node does not actively participate in the transmissions, it would not have enough credits to send its own packets. In [20], users are issued receipts by the central server for their contribution of sensing data in the PS system. The receipts are authenticated by the server but without revealing any sensitive user information; thus, users can gain credits in a privacy-preserving manner.

9.5 Discussion

Based on and beyond the existing solutions for security and privacy-preserving VSNs, there still exist a number of significant security and privacy challenges toward deploying such VSNs. Next, we discuss the challenges and current (yet not complete) efforts toward them.

9.5.1 Inference Attacks

In VC systems, privacy is a main concern due to the location information included in transmitted messages. This has been addressed by pseudonymous authentication; thus, locations in the messages cannot be linked to any real vehicle identity. However, location information and the structure of the roads (i.e., mobility restrictions) make it relatively easy to link messages. An attacker that eavesdrops on all beacons within an area is able to track the vehicles with almost 100% accuracy [50]. In fact, in a few cases, an attacker would be able to eavesdrop on all messages within an area; however, richer data exchanged within VSNs made tracking based on content of messages possible. For example, if a vehicle disseminates a same piece of information during its trip, then it can be easily tracked based on the message content. Although VSN interactions could be strictly based on user relationships leveraging data encryption, a significant amount of VSN interactions are region based (thus, any eavesdropper can read the information). As the VSN applications become popular, more and more data will be exchanged among the vehicles; thus, the privacy of users is at stake.

9.5.2 Identity Management

In OSNs, without any validation on true user identities, creation of fake accounts or identities is possible. Such Sybil-based misbehavior could lead to distribution of false or misleading rumors, so that benign users are affected by the adversaries. VSNs, leveraging strong identification of vehicles, could prevent the creation of fake identities: each registered user has only one long-term identity. However, it is possible that each user equips him or herself with several pseudonymous identities in multiple VSNs. The possibility of interactions among different VSNs could be exploited by an adversary to appear as a Sybil node in the VSNs (e.g., disseminating data about a same event from multiple VSNs). Binding of the pseudonyms of a user together could solve the problem, while user interest could be disclosed in terms of the VSNs one has joined (thus breaching user privacy).

HSM-equipped vehicles are not able to manipulate pseudonym usage. However, considering the heterogeneity of participants in VSNs (e.g., smartphone users), it is unrealistic to assume every node in the VSNs would be equipped with an HSM. This makes sharing of pseudonyms possible. Multiple users can share their pseudonyms (along with corresponding private keys) and appear as Sybil nodes in different areas without being detected. Even though the ownership of pseudonyms is maintained by the security infrastructure and the real identity of the pseudonym owner could be resolved in the case of misbehavior, such an on-demand process does not support real-time detection and prevent the users from sharing or transferring their credentials.

As described in Section 9.4.5, there are cases when user contribution needs to be recorded and incentivized. Leveraging relatively long-lived pseudonyms, users can remain operational within VSNs that applied incentivized schemes. However, keeping credits while using short-term pseudonyms would be a problem, since incentives would be lost once a new pseudonym is used. To maintain the incentives, a central server is needed to record incentives for the pseudonyms of a user. However, such an approach conflicts with the motivation behind pseudonymous authentication: any two pseudonyms of a same user should not be linkable, as this requires the server to identify which pseudonyms correspond to which long-term identities. This leaves a challenge for accumulating credits for user contribution while preserving user privacy.

9.6 Conclusions

In this chapter, we surveyed the state-of-the-art security and privacy architectures and technologies for VC systems, emphasizing security and privacy for VSNs. Beyond transportation safety and efficiency applications that have drawn a lot of attention in VC systems, there is significant and rising interest in V2V interaction for transportation efficiency and infotainment applications, notably, LBS and a gamut of services by mobile providers. While this enriches the VC systems and the user experience, security and privacy concerns are also intensified. This is especially so considering (1) the privacy risk from the exposure of the users to the SPs, and (2) the security risk from the interaction with malicious or selfish, and thus misbehaving, users or infrastructure. We showed existing solutions can in fact evolve and address the VSN-specific challenges, and improve or even accelerate the adoption of VSN applications.

References

1. Nikolaos Alexiou, Marcello Laganà, Stylianos Gisdakis, Mohammad Khodaei, and Panos Papadimitratos. VeSPA: Vehicular security and privacy-preserving architecture. In *ACM HotWiSec*, pp. 19–24, Budapest, Hungary, April 2013.
2. Norbert Bißmeyer, Jonathan Petit, and Kpatcha M Bayarou. CoPRA: Conditional pseudonym resolution algorithm in VANETs. In *IEEE WONS*, pp. 9–16, Banff, Canada, March 2013.
3. Dan Boneh, Xavier Boyen, and Hovav Shacham. Short group signatures. In *Advances in Cryptology—CRYPTO*, pp. 41–55, Santa Barbara, CA, August 2004.
4. Dan Boneh and Hovav Shacham. Group signatures with verifier-local revocation. In *ACM CCS*, pp. 168–177, Washington, DC, October 2004.
5. Jeffrey A. Burke, Deborah Estrin, Mark Hansen, Andrew Parker, Nithya Ramanathan, Sasank Reddy, and Mani B. Srivastava. Participatory sensing. Los Angeles: Center for Embedded Network Sensing, 2006.
6. Giorgio Calandriello, Panos Papadimitratos, Jean-Pierre Hubaux, and Antonio Lioy. On the performance of secure vehicular communication systems. *IEEE Transactions on Dependable and Secure Computing*, 8(6):898–912, 2011.

7. Giorgio Calandriello, Panos Papadimitratos, Jean-Pierre Hubaux, and Antonio Lioy. Efficient and robust pseudonymous authentication in VANET. In *ACM VANET*, pp. 19–28, Montreal, Quebec, September 2007.

8. Car-to-Car Communication Consortium (C2C-CC). http://www.car-2-car.org/.

9. David Chaum and Eugène Van Heyst. Group signatures. In *Advances in Cryptology—EUROCRYPT*, pp. 257–265, Brighton, UK, April 1991.

10. Leucio Antonio Cutillo, Refik Molva, and Thorsten Strufe. Safebook: Feasibility of transitive cooperation for privacy on a decentralized social network. In IEEE WoWMoM, pp. 1–6, Kos, Greece, June 2009.

11. Stefan Dietzel, Julian Gurtler, Rens van der Heijden, and Frank Kargl. Redundancy-based statistical analysis for insider attack detection in VANET aggregation schemes. In *IEEE VNC*, pp. 135–142, Paderborn, Germany, December 2014.

12. John R. Douceur. The sybil attack. In *ACM Peer-to-Peer Systems*, pp. 251–260, London, March 2002.

13. TCITS ETSI. ETSI Technical Specification (TS) 103 097 v1. 1.1-Intelligent Transport Systems (ITS); Security; Security header and certificate formats, 2013.

14. Mohamed Fazeen, Brandon Gozick, Ram Dantu, Moiz Bhukhiya, and Marta C. González. Safe driving using mobile phones. *IEEE Transactions on Intelligent Transportation Systems*, 13(3):1462–1468, 2012.

15. Jon Froehlich, Tawanna Dillahunt, Predrag Klasnja, Jennifer Mankoff, Sunny Consolvo, Beverly Harrison, and James A. Landay. UbiGreen: Investigating a mobile tool for tracking and supporting green transportation habits. In *International Conference on Human Factors in Computing Systems*, pp. 1043–1052, April 2009.

16. Raghu K. Ganti, Nam Pham, Hossein Ahmadi, Saurabh Nangia, and Tarek F. Abdelzaher. GreenGPS: A participatory sensing fuel-efficient maps application. In *ACM MobiSys*, pp. 151–164, San Francisco, June 2010.

17. Raghu K. Ganti, Fan Ye, and Hui Lei. Mobile crowdsensing: Current state and future challenges. *IEEE Communications Magazine*, 49(11):32–39, 2011.

18. Bugra Gedik and Ling Liu. Protecting location privacy with personalized k-anonymity: Architecture and algorithms. *IEEE Transactions on Mobile Computing*, 7(1):1–18, 2008.

19. Gabriel Ghinita, Panos Kalnis, and Spiros Skiadopoulos. Mobihide: A mobile peer-to-peer system for anonymous location-based queries. In *Advances in Spatial and Temporal Databases*, pp. 221–238, Boston, July 2007.

20. Stylianos Gisdakis, Thanassis Giannetsos, and Panos Papadimitratos. Sppear: Security & privacy-preserving architecture for participatory-sensing applications. In *Proceedings of the 2014 ACM Conference on Security and Privacy in Wireless & Mobile Networks*, pp. 39–50, 2014.

21. Stylianos Gisdakis, Thanassis Giannetsos, and Panos Papadimitratos. Shield: A data verification framework for participatory sensing systems. In *ACM WiSec*, pp. 16:1–16:12, New York, June 2015.

22. Stylianos Gisdakis, Marcello Laganà, Thanassis Giannetsos, and Panos Papadimitratos. SEROSA: Service oriented security architecture for vehicular communications. In *IEEE VNC*, pp. 111–118, Boston, December 2013.

23. Stylianos Gisdakis, Vasileios Manolopoulos, Sha Tao, Ana Rusu, and Panos Papadimitratos. Secure and privacy-preserving smartphone-based traffic information systems. *IEEE Transactions on Intelligent Transportation Systems*, 16(3):1428–1438, 2015.

24. Aris Gkoulalas-Divanis, Panos Kalnis, and Vassilios S Verykios. Providing k-anonymity in location based services. *ACM SIGKDD Explorations Newsletter*, 12(1):3–10, 2010.

25. Glenn Greenwald. NSA prism program taps in to user data of Apple, Google and others. Guardian, June 7, 2013, p. 1.

26. IEEE P1609.2/D12. Draft standard for wireless access in vehicular environments. Piscataway, NJ: Institute of Electrical and Electronics Engineers, January 2012.

27. Hongyu Jin and Panos Papadimitratos. Resilient collaborative privacy for location-based services. In *Nordic Conference on Secure IT Systems*, pp. 47–63, Stockholm, October 2015.

28. Apu Kapadia, David Kotz, and Nikos Triandopoulos. Opportunistic sensing: Security challenges for the new paradigm. In *COMSNETS*, pp. 1–10, Bangalore, January 2009.

29. Mohammad Khodaei, Hongyu Jin, and Panos Papadimitratos. Towards deploying a scalable & robust vehicular identity and credential management infrastructure. In *IEEE VNC*, pp. 33–40, Paderborn, Germany, December 2014.

30. Mohammad Khodaei and Panos Papadimitratos. The key to intelligent transportation: Identity and credential management in vehicular communication systems. *IEEE Vehicular Technology Magazine*, 10(4):63–69, 2015.

31. Tim Leinmüller, Levente Buttyan, Jean-Pierre Hubaux, Frank Kargl, Rainer Kroh, Panos Papadimitratos, Maxim Raya, and Elmar Schoch. SeVeCom—Secure vehicle communication. In *IST Mobile and Wireless Communication Summit*, Mykonos, Greece, June 2006.

32. Xiaodong Lin, Xiaoting Sun, Pin Han Ho, and Xuemin Shen. GSIS: A secure and privacy-preserving protocol for vehicular communications. *IEEE Transactions on Vehicular Technology*, 56(6):3442–3456, 2007.

33. Rongxing Lu, Xiaodong Lin, Haojin Zhu, Pin-Han Ho, and Xuemin Shen. ECPP: Efficient conditional privacy preservation protocol for secure vehicular communications. In *IEEE INFOCOM*, pp. 1903–1911, Phoenix, AZ, April 2008.

34. Guilherme Maia, Andre L.L. Aquino, Aline Viana, Azzedine Boukerche, and Antonio A.F. Loureiro. HyDi: A hybrid data dissemination protocol for highway scenarios in vehicular ad hoc networks. In *ACM DIVANet*, pp. 115–122, Paphos, Cyprus Island, October 2012.

35. Vasileios Manolopoulos, Panos Papadimitratos, Sha Tao, and Ana Rusu. Securing smartphone based ITS. In *ITS Telecommunications*, pp. 201–206, St. Petersburg, Russia, August 2011.

36. Ghita Mezzour, Adrian Perrig, Virgil Gligor, and Panos Papadimitratos. Privacy-preserving relationship path discovery in social networks. In *International Conference on Cryptology and Network Security*, pp. 189–208, Kanazawa, Japan, December 2009.

37. Mohamed F. Mokbel, Chi-Yin Chow, and Walid G. Aref. The new Casper: Query processing for location services without compromising privacy. In *Proceedings of the 32nd International Conference on Very Large Data Bases*, pp. 763–774, Seoul, September 2006.

38. Panos Papadimitratos, Giorgio Calandriello, Jean-Pierre Hubaux, and Antonio Lioy. Impact of vehicular communication security on transportation safety. In *IEEE INFOCOM MOVE*, pp. 1–6, Phoenix, AZ, April 2008.

39. Panos Papadimitratos, Levente Buttyan, Tamas Holczer, Elmar Schoch, Julien Freudiger, Maxim Raya, Zhendong Ma, Frank Kargl, Antonio Kung, and Jean-Pierre Hubaux. Secure vehicular communication systems: Design and architecture. *IEEE Communications Magazine*, 46(11):100–109, 2008.

40. Panos Papadimitratos, Levente Buttyan, Jean-Pierre Hubaux, Frank Kargl, Antonio Kung, and Maxim Raya. Architecture for secure and private vehicular communications. In *ITST*, pp. 1–6, Sophia Antipolis, France, June 2007.

41. Panos Papadimitratos, Virgil Gligor, and Jean-Pierre Hubaux. Securing vehicular communications—Assumptions, requirements, and principles. In *ESCAR*, pp. 5–14, Berlin, November 2006.

42. Maxim Raya, Panos Papadimitratos, Virgil Gligor, and Jean-Pierre Hubaux. On data-centric trust establishment in ephemeral ad hoc networks. In *IEEE INFOCOM*, pp. 1912–1920, Phoenix, AZ, April 2008.

43. Krishna Sampigethaya, Mingyan Li, Leping Huang, and Radha Poovendran. AMOEBA: Robust location privacy scheme for VANET. *IEEE Journal on Selected Areas in Communications*, 25(8):1569–1589, 2007.

44. Florian Schaub, Frank Kargl, Zhendong Ma, and Michael Weber. V-tokens for conditional pseudonymity in VANETs. In *IEEE WCNC*, pp. 1–6, Sydney, NSW, 2010.

45. Reza Shokri, Georgios Theodorakopoulos, Panos Papadimitratos, Ehsan Kazemi, and Jean-Pierre Hubaux. Hiding in the mobile crowd: Location privacy through collaboration. *IEEE Transactions on Dependable and Secure Computing*, 11(3):266–279, 2014.

46. Stephen Smaldone, Chetan Tonde, Vancheswaran K. Ananthanarayanan, Ahmed Elgammal, and Liviu Iftode. The cyber-physical bike: A step towards safer green transportation. In *HotMobile*, pp. 56–61, Phoenix, AZ, March 2011.

47. Jan Peter Stotz, Norbert Bißmeyer, Frank Kargl, Stefan Dietzel, Panos Papadimitratos, and Christian Schleiffer. Security requirements of vehicle security architecture, PRESERVE—Deliverable 1.1. Seventh Framework Programme, June 2011.

48. Ahren Studer, Elaine Shi, Fan Bai, and Adrian Perrig. Tacking together efficient authentication, revocation, and privacy in VANETs. In *IEEE SECON*, pp. 1–9, Rome, June 2009.

49. Akriti Verma, Deepak Kshirsagar, and Sana Khan. Privacy and security: Online social networking. *International Journal of Advanced Computer Research*, 3(8):310–315, 2013.

50. Björn Wiedersheim, Zhendong Ma, Frank Kargl, and Panos Papadimitratos. Privacy in inter-vehicular networks: Why simple pseudonym change is not enough. In *IEEE WONS*, pp. 176–183, Kranjska Gora, Slovenia, February 2010.

51. Jorge Zaldivar, Carlos T. Calafate, Juan Carlos Cano, and Pietro Manzoni. Providing accident detection in vehicular networks through OBD-II devices and android-based smartphones. In *IEEE Conference on Local Computer Networks*, pp. 813–819, Bonn, Germany, October 2011.

52. Jun Zhou, Xiaolei Dong, Zhen-Fu Cao, and Athanasios Vasilakos. Secure and privacy preserving protocol for cloud-based vehicular DTNs. *IEEE Transactions on Information Forensics and Security*, 10(6):1299–1314, 2015.

Index

A

ABSM, 107, 120
Aggregated local mobility (ALM) algorithm, 5
AID, 107, 123
Airbnb, 145
ALM algorithm, *see* Aggregated local mobility (ALM) algorithm
AMACAD, 6
ASPIRE clustering algorithm, 6

B

Bee colony-enlivened interest-based forwarding (BEEINFO), 7, 63
Big Data, 20
Bonfyr, 86
Broadcast storm problem, 104
Broadcast suppression
    delay-based solution, 110–111
    probabilistic-based solution, 113–114
Bundle protocol, 88–89

C

CAMs, *see* Cooperative awareness messages (CAMs)
Carbon emission model, 68–70
CenceMe, 27
Centrality
    betweenness, 45–46
    bridging, 46
    closeness, 45
    social path duration, 46
Centralized social sensing, 26–28
    applications, 27–29
    vehicle-to-infrastructure communications, 27
CH, *see* Cluster head (CH)
CLA, *see* Convergence layer adapter (CLA)
Click router, 94
Cluster head (CH), 4
    ALM, 5

ASPIRE, 6
ES-Cl, 5
HCA, 6
SRB, 7
Clustering
    of nodes, 4
    social vehicles, 6–7
    vehicular networks, 5–6
Clustering coefficient solution
    delay-based solution, 112
    probabilistic-based solution, 114–116
Cluster member (CM), 5
Cluster relay (CR) nodes, 6
CM, *see* Cluster member (CM)
CMEM, *see* Comprehensive modal emissions model (CMEM)
Collaborative economy, 145
Communitywide metrics
    number of clusters, 47
    number of social groups, 47
Comprehensive modal emissions model (CMEM), 69
Convergence layer adapter (CLA), 88, 90–92
Cooperative awareness messages (CAMs), 10
CR nodes, *see* Cluster relay (CR) nodes
Crowdsourcing
    benefits of, 131
    challenges, 136–137
    definition, 132
    in VANETs, 131
    for vehicular social networks, 134–136

D

Data dissemination, 104
    ABSM, 107
    AID, 107
    definition, 105
    DRIVE, 107
    DV-CAST, 106
    SimBet, 105–106